T0255092

# Lecture Notes in Mathematics

Volume 2336

This series reports on new developments in all areas of mathematics and their applications - quickly, informally and at a high level. Mathematical texts analysing new developments in modelling and numerical simulation are welcome. The type of material considered for publication includes:

1. Research monographs
2. Lectures on a new field or presentations of a new angle in a classical field
3. Summer schools and intensive courses on topics of current research.

Texts which are out of print but still in demand may also be considered if they fall within these categories. The timeliness of a manuscript is sometimes more important than its form, which may be preliminary or tentative.

*Titles from this series are indexed by Scopus, Web of Science, Mathematical Reviews, and zbMATH.*

Olga Gil-Medrano

# The Volume of Vector Fields on Riemannian Manifolds

## Main Results and Open Problems

Olga Gil-Medrano
Department of Mathematics
University of Valencia
Valencia, Spain

ISSN 0075-8434 ISSN 1617-9692 (electronic)
Lecture Notes in Mathematics
ISBN 978-3-031-36856-1 ISBN 978-3-031-36857-8 (eBook)
https://doi.org/10.1007/978-3-031-36857-8

Mathematics Subject Classification: 53C20, 53C42, 58E15 , 53C24

This work was supported by Ministerio de Ciencia e Innovación (Project PID2019-105019GB-C21)



This Springer imprint is published by the registered company Springer Nature Switzerland AG
The registered company address is: Gewerbestrasse 11, 6330 Cham, Switzerland

# Preface

The purpose of this monograph is to provide a comprehensive overview of the volume of vector fields on a Riemannian manifold, including the proof of the more relevant results obtained from the time it was first introduced in 1986. Our aim is also to exhibit some of the many open problems, with special emphasis in those that involve the spheres with their usual metric in the hope of encouraging further research.

We have participated actively in the development of the subject, and after putting it at a distance during a period, we have proceeded to a complete review of the results and chosen the most direct way to present them as far as the approach that seems to us the most effective. This has required a deep revision of the definitions, statements and demonstrations which are mostly different from those that appeared in the papers cited. The text also contains some unpublished results. The primary audience of the book consists of researchers and PhD students interested in geometric analysis problems.

All the chapters end with a Notes section containing complementary information that is relevant to comprehend the subject in deep but whose reading can be omitted without impairing the understanding of the rest. The Notes as a unit can also be seen as a survey that updates our previous surveys [40] and [41].

I want to express my gratitude to the colleagues with whom I have worked in this field: Vincent Borrelli, Fabiano Brito, Carmelo González-Dávila, Lieven Vanhecke and especially my two former students Ana Hurtado and Elisa Llinares-Fuster. It was a privilege to share so many hours with all of them. The work would not have been possible without the support of the members of the Geometric Analysis research group of the University of Valencia; I am very grateful to all of them and in particular to Vicente Miquel for including me, even after my retirement, in the research project that he leads and for his constant encouragement. Thanks to the work of the editors of Springer Verlag and the reviewers, the version that reaches the reader contains notable improvements with respect to the initial manuscript. My sincere thanks.

## Funding Acknowledgements

The author was supported by grant PID2019-105019GB-C21 funded by MCIN/AEI/ 10.13039/501100011033 and by ERDF (European Regional Development Fund).

Valencia, Spain
May 2023

Olga Gil-Medrano

# Contents

# Chapter 1
# Introduction

A vector field on a differentiable manifold is a section of the manifold in its tangent bundle and is therefore an injective immersion. The image of the vector field is an embedded submanifold of the tangent manifold. A Riemannian metric on the manifold fully determines a natural Riemannian metric on the tangent manifold that was first defined by Sasaki in [86]. The relationship between the geometry of both manifolds is well known.

In the late nineties, we introduced the study of the condition for a vector field to determine a minimal submanifold of its tangent bundle endowed with the Sasaki metric, both for the interest of this concept in itself—because it is the natural analog to that of being a minimal graph in the event that the manifold is an open subset of the Euclidean space—and especially to better understand and advance in the solution of the problem that Gluck and Ziller had raised a few years before.

In their 1986 paper [52], the aforementioned authors proposed the idea that vector fields realising the minimum volume should be recognisable by their good geometric properties and showed that this is true for three-dimensional spheres. To be more precise, since a normalisation is needed because the identically zero vector field is the one that always has the smallest volume, the question posed in [52] is: What are the unit vector fields of the odd-dimensional spheres that have the minimum volume? They show that, in the case of the sphere of dimension three, these are exactly the vector fields tangent to fibers of the Hopf fibration. Surprisingly, despite the number of articles that have been published on the subject, the problem of determining the infimum of the volume of smooth unit vector fields on the spheres of higher odd dimension remains open.

Given a Riemannian manifold, to find the infimum of the volume functional and the minimising unit vector fields, if they exist, one can first search among the critical points of the functional defined in the space of smooth unit vector fields. In the joint work with E. Llinares-Fuster [47] we compute the first variation and have shown that a unit vector field is a critical point if and only if it determines a minimal submanifold of the unit tangent bundle, i. e. if and only if its mean curvature

 
O. Gil-Medrano, *The Volume of Vector Fields on Riemannian Manifolds*,
Lecture Notes in Mathematics 2336, https://doi.org/10.1007/978-3-031-36857-8_1

vanishes. Using this variational approach we showed that, for the odd dimensional spheres of any radius, the Hopf vector fields define minimal submanifolds of the unit tangent bundle; the question of whether they are the only smooth unit vector fields on the sphere with this property is open even for dimension three

The study of the stability of Hopf vector fields, as critical points of the volume restricted to unit vector fields, developed in our joint work with E. Llinares-Fuster [46] and with V. Borrelli [11], revealed the sensitivity of the problem with respect to the variation of the radius of the sphere. The nature of this question, which depends on the geometry of the Sasaki metric and not on the particular case of the spheres, has led us to consider thorough this monograph sections of constant length and not only unit sections; this approach makes the results for spheres of any radius more transparent—shown here instead for vector fields of any length—and has the advantage of being suitable for any Riemannian manifold.

The lower bound of the volume of smooth unit vector fields for spheres of any odd-dimension indicated in [52] was generalised by F. G. B. Brito, P. M. Chacón and A. M. Naveira to manifolds of constant curvature in [21] where they proved that, for dimension greater than three, the bound cannot be attained by any smooth vector field, but is the value of the volume of a family of vector fields with two singularities.

The role of vector fields with singularities in the possible solution of the problem had already been highlighted by S. Pedersen in [78] who exhibited a family of unit vector fields with one singularity, whose volume can be approximated by smooth unit vector fields, which allowed her to show that the Hopf vector fields cannot be the volume minimisers of the volume of unit vector field for spheres of odd-dimensions greater than three. In our joint work with V. Borrelli [12] we have shown that these vector fields play a central role in the solution of the corresponding problem on the 2-dimensional sphere, since they are area minimising among the vector fields of constant length as regular as possible.

In conclusion, many results have been obtained advancing towards the solution of the problems described above which, as usual, have raised new questions; this book is devoted to exposing both results and open questions.

We proceed below to describe in detail the content of the four chapters into which the book is divided.

The process to build the Sasaki metric $g^S$ on the tangent manifold is quite simple: at each point $\sigma \in TM$ we consider the decomposition of $T_\sigma TM$ into two subspaces, the vertical one being the tangent to the fibers and the horizontal one, which is the complementary to the vertical determined by the Levi-Civita connection of $g$. The metric $g^S$ in $T_\sigma TM$ is defined in such a way that these two subspaces are orthogonal, that $g^S = g$ when restricted to the tangent space to the vertical subspace and that on the horizontal subspace it makes the map $\pi : (TM, g^S) \to (M, g)$ a Riemannian submersion.

The Sasaki metric of the tangent bundle has a natural extension to a Riemannian metric on any tensor bundle $\pi : P \to M$ of a Riemannian manifold $(M, g)$ that we describe in Sect. 2.1. The unit tensor bundles $S^1P$, or more generally the sphere bundles of tensors of constant length $S^rP$, are considered with the induced metric. The Sasaki metric is not the only metric on the tensor bundles that can be constructed with $g$ as the only data, but it's the most natural one since, for example, with this metric the unit tangent bundle of $S^2$ is isometric to the projective space (W. Klingenber and S. Sasaki, [70]) and in general $T^1S^n$ is isometric to the Stiefel manifold of orthonormal two-frames of $\mathbf{R}^{n+1}$ (H. Gluck and W. Ziller, [52]).

For any smooth section $\sigma : M \to (P, g^S)$, or $\sigma : M \to (S^r P, g^S)$, we study the extrinsic geometry of the submanifold $\sigma(M)$ calculating the second fundamental form and the mean curvature vector field. As a corollary, we obtain the condition that a section must satisfy for $\sigma(M)$ to be a totally geodesic or minimal submanifold.

Section 2.2 is devoted to minimal and $r$-minimal sections, which are defined as those $\sigma$ such that $\sigma(M)$ is a minimal submanifold of $P$ or $S^r P$, respectively. We transform the minimality condition into a PDE on $\sigma$ and its first and second covariant derivatives, $\nabla\sigma$ and $\nabla^2\sigma$, thus characterising the minimal sections as the solutions of $\nabla^* K_\sigma = 0$ and the $r$-minimal sections as the solutions of

$$\nabla^* K_\sigma = \frac{1}{r^2} g(K_\sigma, \nabla\sigma)\sigma,$$

where $\nabla^*$ represents the divergence operator, $K_\sigma$ is the section of $P \otimes T^*M$ that acting on vector fields gives

$$K_\sigma(Y) = f_\sigma \nabla_{L_\sigma^{-1}(Y)}\sigma,$$

with $L_\sigma$ being the endomorphism field defined by

$$g(L_\sigma(X), Y) = \sigma^* g^S(X, Y) = g(X, Y) + g(\nabla_X\sigma, \nabla_Y\sigma)$$

and $f_\sigma = \sqrt{\det L_\sigma}$.

A first result is that if $M$ is closed (compact, without boundary) any minimal section $\sigma$ must be parallel, i. e. $\nabla\sigma = 0$.

From here we focus on the main topic of the monograph, the Volume of vector fields. For a vector field $V$ on a Riemannian manifold $(M, g)$ of dimension $n$, it is defined as the $n$-dimensional volume of $V(M) \subset (TM, g^S)$, or equivalently, as the volume of the manifold $M$ endowed with the induced metric $V^* g^S$, it is given by

$$\mathrm{Vol}(V) = \int_M \sqrt{\det(\mathrm{Id} + (\nabla V)^t \circ \nabla V)}\,\mathrm{dv}_g.$$

The condition of minimality (resp. $r$-minimality) exposed above turns out to be the Euler-Lagrange equation of the variational problem associated to the Volume functional, defined in the space of smooth vector fields (resp. smooth vector fields of length $r$). When $M$ is closed, the equation characterises the critical points. In Sects. 2.3 and 2.4 we study the first and second variation of the Volume functional, respectively. We do it following the lines of our joint papers with E. Llinares-Fuster

[46] and [47], the difference is that we have considered here not only the restriction of the functional to unit vector fields but also to vector fields of any constant length.

To finish the chapter, in Sect. 2.5 we dissect in detail the particular case of the manifolds of dimension 2.

Chapter 3 is devoted to the study of the smooth $r$-minimal vector field on the odd-dimensional spheres. The unit vector fields tangent to the fibers of the Hopf fibration $\pi : S^{2m+1} \to \mathbf{C}P^{m+1}$ are given by $H = iN$, where $N$ is the outwards unit normal of the sphere. More generally, if $J$ is a complex structure of $\mathbf{R}^{2m+2}$ the vector field of the form $JN$ is known as a Hopf field; here we will also call Hopf vector fields those of the form $rJN$ for a positive real number $r$.

We show that $rH$ is $r$-minimal for all $r$ and here the first interesting open problem appears: to determine if they are the only smooth $r$-minimal vector fields defined on the entire sphere. So far they are the only known examples. We give in Sect. 3.1 some evidence that leads us to conjecture that every smooth minimal vector field of constant length in $S^{2m+1}$ must be a Hopf vector field.

Computing the Hessian of the Volume functional at a minimal vector field on a general manifold is extremely complicated, but at the Hopf vector fields of spheres of odd dimension it is possible to show that it admits the expression

$$(Hess\mathrm{Vol})_{H^r}(A) = (1+r^2)^{m-2} \int_{S^{2m+1}} \left(-2m\|A\|^2 + \|\nabla A\|^2 + r^2\|JA + \nabla_H A\|^2\right) \mathrm{dv}_g$$

which allows us to show in Sect. 3.2 that for $r < \sqrt{2m-3}$ Hopf vector fields are unstable. We use the same proof here as in [46] where we exhibit vector fields orthogonal to $H$ such that the Hessian in its direction is negative.

The main result of the chapter is that for $r \geq \sqrt{2m-3}$ the Hopf vector fields are stable. This was shown in our joint paper with V. Borrelli [11] and an important idea of the proof was to write the Hessian of the volume at $rH$, in the direction of a vector field $A$, in terms of the Fourier series of the restriction of $A$ to the fibers of the Hopf fibration. More precisely, if we define on the sphere the $\mathbf{R}^{2m+2}$ valued smooth maps

$$A_k(p) = \frac{1}{2\pi} \int_0^{2\pi} e^{-ikt} A(e^{it} p) dt,$$

we show that $(Hess\mathrm{Vol})_{rH}(A) = \sum_{k \in \mathbf{Z}} (Hess\mathrm{Vol})_{rH}(A_k)$.

The next step is to show that, under the condition assumed on $r$, all the terms of the series are positive. In the proof given in the book, we have simplified the arguments for this step by estimating the Hessian by the first and second eigenvalues of the rough Laplacian, using the characterisation of the eigenvalues via Rayleigh quotients, which for the sphere gives that

$$\int_{S^{2m+1}} \|\nabla A\|^2 \mathrm{dv}_g \geq \int_{S^{2m+1}} \|A\|^2 \mathrm{dv}_g$$

and, more importantly, that if we restrict to vector fields whose coordinates have zero integral then

$$\int_{S^{2m+1}} \|\nabla A\|^2 \mathrm{dv}_g \geq 2m \int_{S^{2m+1}} \|A\|^2 \mathrm{dv}_g.$$

The condition of having coordinates of zero integral is equivalent to being $L_2$-orthogonal to the first eigenspace of the rough Laplacian. So, we just need to show that $(Hess\mathrm{Vol})_{rH}(A_0)$ is non-negative and we do this by finding an estimate of the norm $L_2$ of $\nabla A$ for vector fields that are orthogonal to $H$. This simplification of the proof will reveal its importance in the next section, devoted to space forms of positive curvature.

A main property of Hopf vector fields is that they are the constant length Killing vector fields of the odd-dimensional spheres. By extension, we define the Hopf vector fields on space forms of positive curvature $M = S^{2m+1}/G$ as the Killing vector fields of constant length on $M$, they can also be described as the projections to $M$ of the Hopf vector fields of $S^{2m+1}$. In [47] we have shown that they are $r$-minimal; to be precise, every Killing vector field $V$ of length $r > 0$ on a $n$-dimensional manifold $M$ of constant curvature $c$ is $r$-minimal with volume $\mathrm{Vol}(V) = (1 + cr^2)^{\frac{n-1}{2}} \mathrm{vol}(M)$.

After showing that for all the quotients except for the sphere itself, the first eigenvalue of the rough Laplacian is $2m$, with eigenspace consisting of all Killing vector fields, we show in a simple way an important result obtained with a different method by V. Borrelli and H. Zoubir in [13]: Hopf vector fields of any constant length defined on $M = S^{2m+1}/G$, with $G \neq \{\mathrm{Id}\}$, are stable.

The Notes Section of this chapter can be considered as a survey of the many examples of minimal vector fields, and the few examples of other tensor sections, obtained by several authors from the year 2000 to now.

The second part of the book, Chaps. 4 and 5, is devoted mainly to studying the following questions: What geometric meaning does the infimum of the Volume functional on vector fields of constant length have? Can we compute this infimum explicitly for some Riemannian manifolds? Is this infimum reached by some smooth vector field? Can we determine exactly the fields that minimise the volume?

For a Riemannian manifold with finite volume, the real numbers

$$\mathcal{V}(M, r) = \inf\{\mathrm{Vol}(V)\ ;\ V \in \Gamma^\infty(T^r(M))\},$$

for $r > 0$, are bounded below by the $\mathrm{vol}(M)$ and the bound is attained only by parallel vector fields. But the existence of a parallel $V$, other than $V \equiv 0$, implies the vanishing of the sectional curvature of the planes containing $V$ and therefore it is natural to expect that, for manifolds without non-vanishing parallel vector fields, the value $\mathcal{V}(M, r)$ encodes some geometrical information. On the other hand, if the infimum is attained for a given $V$, it's also natural to expect its covariant derivative to have some properties.

One of the main results in Chap. 4 is that on a complete 3-dimensional manifold of constant curvature $c > 0$ (a spherical space form for short) $\mathscr{V}(M,r) = (1 + cr^2)\mathrm{vol}(M)$ and that the minimisers are exactly the Hopf vector fields; or in other words, the volume-minimising vector fields are exactly those for which $\nabla V$ is skew-symmetric.

This Theorem is a generalisation of the main result of [52] cited above, which was valid for unit vector fields on the unit sphere. The proof we show in Sect. 4.1 is not an extension of the proof in [52] where the authors used the method of calibrations, but it follows from completely different arguments developed in our recent paper [43]. To use the calibration method, the authors take advantage of the fact that for all the unit vector fields of $S^3$ the submanifolds $V(S^3) \subset T^1S^3$ are in the same homology class. As H. Gluck and W. Ziller point out,

> The drawn back to using the method of calibrated geometries for this problem is that we must prove a little more than we want: the $V_H$ have minimum volume among all 3-manifolds in the same homology class in $T^1S^3$... As a result, the method will fail on the five-sphere.....

To prove the generalisation stated above, we use the properties of the first eigenvalues of the rough Laplacian acting on vector fields and the fact that if $V$ is volume-minimiser then it must be $r$-minimal. These arguments do not involve the homology class.

Although the chapter is dedicated to spheres with the round metric and their quotients, we have included in Sect. 4.2 a result about the volume-minimising vector fields on $S^3$ endowed with the Berger metrics. This one-parameter family of metrics $g_\mu$, for $\mu > 0$, on the odd-dimensional spheres, is defined by deforming the usual metric by a factor $\mu$ in the direction of $H$ and includes the usual metric taking $\mu = 1$. We will show that the Hopf vector fields are exactly the minimisers of the volume if and only if $\mu \leq 1$. This result, which we proved in our joint paper with A. Hurtado [45], was obtained for $\mu < 1$ with a completely different proof to that of $\mu = 1$, the main ingredient being the use of the identification of $\mathbf{R}^4$ with the quaternions. The existence of minimising unit vector fields when $\mu > 1$ is still an open problem.

In Sect. 4.3, manifolds of constant curvature of any odd dimension are considered for which F. G. B. Brito, P. M. Chacón and A. M. Naveira computed in [21] a lower bound of the volume of unit vector fields in terms of the curvature. The proof we reproduce here uses their main central idea but we have refined the linear algebra technicalities to make it shorter. To explain the main idea in [21] we need to express the Volume functional in terms of the symmetric polynomials $\sigma_k$ of the endomorphism field $(\nabla V)^t \nabla V$

$$\mathrm{Vol}(V) = \int_M \sqrt{1 + \sigma_1((\nabla V)^t \nabla V) + \cdots + \sigma_{n-1}((\nabla V)^t \nabla V)}\, \mathrm{dv}_g.$$

For a manifold of constant curvature, if $P$ represents the restriction of $\nabla V$ to $V^\perp$, the value of the integral of $\sigma_{2k}(P)$ is known and turns out to be independent of the unit vector field as was proved by F. G. B. Brito, R. Langevin and H. Rosenberg in [20].

Thus, the effort is devoted to estimating $\mathrm{Vol}(V)$ by the integrals of the symmetric polynomials of $P$ of even degree.

As noted in [21], for $S^{2m+1}$ the lower bound is the value of the volume of vector fields that are tangent to radial geodesics issuing from any given point, which are smoothly defined in the sphere minus two antipodal points. But as they have shown, the bound cannot be the volume of a smooth vector field defined in the entire sphere, unless $m = 1$. The result suggests the importance of studying vector fields of constant length defined on the spheres minus a set of isolated points.

In Chap. 5 we first consider radial vector fields and show that they are $r$-minimal. In Sect. 5.2 we concentrate on the vector fields that are obtained by the parallel transport of a given vector $v \neq 0$ tangent at a point $p$, along the radial geodesics issuing from $p$; they will be called parallel transport vector fields and they are smooth on the sphere minus the point $-p$. In the particular case of a unit $v$, these vector fields were considered by S. Pedersen in [78] to show that for $m > 1$ the unit Hopf vector fields are not volume-minimising. One of the steps for the proof was to compute the volume of these vector fields; here we compute the volume of the parallel transport vector fields of any constant length as in [11].

The second step was to show that the volume of the parallel transport vector fields can be approximated by the volume of smooth unit vector fields. Here we give a detailed proof of this fact by constructing for each unit vector field $V$ a variation of vector fields $V_t$ which converges to a parallel transport vector field both when $t$ goes to $\infty$ and when $t$ goes to $-\infty$. In the case of $V = H$ the variational vector fields are precisely those causing the instability results of Chap. 3. This construction is unpublished.

It was shown in [78] that if $P$ is a unit parallel transport vector field of $S^n$ then $P(S^n)$ is a minimal submanifold of the Stiefel manifold known as a Pontryagin cycle. We prove here that, with the exception of $S^2$, parallel transport vector fields are $r$-minimal only for $r = 1$, this unpublished result is to our knowledge the only example of vector fields that are minimal only for a given value of the length.

We also show that for $S^2$ these vector fields are $r$-minimal for all $r$; this fact was previously shown in our joint work with V. Borrelli [12] by a different method. There we proved that the sphere bundle $T^r S^2$ with the Sasaki metric is homothetic to a projective space obtained as a quotient of a Berger sphere, after which we verified that $P(S^2)$ is a projective plane obtained as the quotient of an equatorial sphere of $S^3$. We have shown that they are minimal for all $r$ (totally geodesic for the particular case $r = 1$).

To be more precise, the projective plane mentioned above is obtained as the closure of $P(S^2)$. According to the definition in [12], they are vector fields of constant length without boundary, a concept that describes the idea of being as regular as a section of $T^r S^2$ can be. For $r = 1$ parallel transport vector field are the minimisers of the area among the vector fields without boundary as a consequence of the corresponding minimising property of the totally geodesic projective planes of the projective space, but the proof for the general case was more complicated because a similar result for equatorial projective planes of the Berger projective spaces was not available at the time. Since we have obtained this extension later in [42], the proof that we give in Sect. 5.4 is much simpler.

Throughout the text we have highlighted the open problems that we consider more interesting. The central one, after more than 35 years since its proposal in [52], is to determine the infimum of the volume of unit vector fields on the spheres of odd dimension greater or equal 5 and to know if the value is attained. In view of the results presented in the book it's natural to extend the problem to vector fields of constant length on compact manifolds of constant positive curvature. In Sect. 5.3 we summarise the most relevant results, those that lead us to conjecture that for manifolds of constant positive curvature, other than the sphere, the Hopf vector fields are the only minimisers.

# Chapter 2
# Minimal Sections of Tensor Bundles

The total space of a tensor bundle $\pi : P \to M$ over a Riemannian manifold $(M, g)$ can be endowed in a natural way with a Riemannian metric that generalises the metric defined by T. Sasaki in [88] for the tangent bundle. This metric and its restriction to the unit sphere bundle has been widely studied; we will consider more generally $r$-sphere bundles consisting of elements of $P$ of length $r > 0$.

A section $\sigma$ of a bundle determines an injective immersion of $M$ into the total space of the bundle and the aim of the first two sections of this chapter is to compute the second fundamental form and the mean curvature vector field of $\sigma$ so that we can establish the condition for its vanishing, that is to say the condition for $\sigma(M)$ to be a minimal submanifold. Although the subject of the book is the study of minimal vector fields we will establish the basic ideas and definitions in the more general setting of tensor bundles.

The well known relation between the vanishing of the mean curvature vector field of an immersion and the variation of the volume reads as follows: An immersion $\varphi : M \to (N, h)$ of an $n$-dimensional manifold $M$ into a Riemannian manifold $(N, h)$ is minimal if and only if it is a critical point of the $n$-dimensional volume functional for variations with compact support. The volume functional is defined in the space of immersions of $M$ into $N$ and assigns to each $\varphi$ the $n$-dimensional volume of the image $\varphi(M) \subset (N, h)$, or equivalently the volume of $M$ endowed with the induced metric $(M, \varphi^* h)$.

In the particular case when $(N, h)$ is a tensor bundle over $M$ and $\varphi$ is a section, one can consider also when $\varphi$ is a critical point of the volume functional restricted to variations by sections. This was in fact the first approach to the theory of minimal vector fields in our work [38], see our joint paper with E. Llinares-Fuster [47], where we computed the Euler-Lagrange Equations of this variational problem and showed that a critical point of the volume restricted to vector fields must be a minimal submanifold of the tangent bundle. This variational point of view will be developed in Sect. 2.3 and the second variation of the volume of vector fields will



O. Gil-Medrano, *The Volume of Vector Fields on Riemannian Manifolds*,
Lecture Notes in Mathematics 2336, https://doi.org/10.1007/978-3-031-36857-8_2

we computed in Sect. 2.4. The last section is devoted to the particular case of 2-dimensional manifolds.

## 2.1 Geometry of the Submanifold Determined by a Section of a Tensor Bundle

By a tensor bundle $\pi : P \to M$ we mean a vector bundle such that the fiber at each point $P_x$ is a vector space consisting of tensors of the tangent space $T_x M$. That is, $P_x = (T_x M)_{(r,s)}$ for some $(r, s) \in \mathbf{Z} \times \mathbf{Z}$. In this case, the only data needed to construct a Riemannian metric on $P$ is a Riemannian metric $g$ on $M$. This metric generalises the Sasaki metric of the tangent bundle, (see the paper by Dombrowski [33] and the book by Blair [4], Chapter 9, for example). It will be denoted by $g^S$ and, for $\sigma \in P$, $\xi_1, \xi_2 \in T_\sigma P$, it is defined by

$$g^S(\xi_1, \xi_2) = g(\pi_*(\xi_1), \pi_*(\xi_2)) + g(K(\xi_1), K(\xi_2)),$$

where $K : TP \to P$ is the connection map of the connection on $P$ induced by the Levi Civita connection of $g$; we have represented by $g$ not only the metric on $M$ but also the fibre metric induced on each fibre.

We recall that an element $\xi \in T_\sigma P$ is said to be *vertical* if $\pi_*(\xi) = 0$ and it is said to be *horizontal* if $K(\xi) = 0$.

*Remark* To define a metric generalising the Sasaki metric on the total space of any vector bundle over a Riemannian manifold we only need Riemannian metrics on the base manifold and on the fibres jointly with a connection on the bundle. The metric on $P$ is then defined in such a way that the horizontal distribution determined by the connexion becomes the distribution orthogonal to the fibers. Many of the results in the first section admit a natural generalisation to this more general setting, although we are not going to consider here these more general metrics.

To each smooth section $\sigma$ of $\pi : P \to M$ we can associate a vertical vector field on $P$, denoted by $\sigma^{ver}$ and for each vector field $X$ on $M$, we can define a vector field $X^{hor}$ in $P$, known as its horizontal lift. In a similar way as it has been done by O. Kowalski in [72] for the tangent bundle, we obtain for the Lie brackets the following expressions, for $X, Y \in \Gamma^\infty(TM)$ and $\sigma, \eta \in \Gamma^\infty(P)$:

$$\begin{cases} [X^{hor}, Y^{hor}] \circ \sigma = [X, Y]^{hor} \circ \sigma + (R(X, Y)\sigma)^{ver} \circ \sigma, \\ [X^{hor}, \eta^{ver}] \circ \sigma \ = (\nabla_X \eta)^{ver} \circ \sigma, \\ [\mu^{ver}, \eta^{ver}] \circ \sigma \ = 0. \end{cases} \tag{2.1}$$

Here $\nabla$ is the covariant derivative of the Levi-Civita connection determined by $g$ and $\nabla_X \sigma$ is the usual extension of the derivation $\nabla_X$ of $\Gamma^\infty(TM)$ to the space of tensor fields $\Gamma^\infty(P)$. For $X, Y \in \Gamma^\infty(TM)$, the curvature operator $R(X, Y) : \Gamma^\infty(P) \to$

$\Gamma^\infty(P)$ is

$$R(X,Y)\sigma = \nabla_{[X,Y]}\sigma - \nabla_X\nabla_Y\sigma + \nabla_Y\nabla_X\sigma$$

and then, if we put

$$(\nabla^2\sigma)(X,Y) = \nabla_X\nabla_Y\sigma - \nabla_{(\nabla_X Y)}\sigma,$$

we have

$$R(X,Y)\sigma = (\nabla^2\sigma)(Y,X) - (\nabla^2\sigma)(X,Y).$$

Using (2.1) and Koszul's formula, we get for the Levi Civita connection of $g^S$:

$$\begin{cases} (\nabla^S_{\mu^{ver}}\eta^{ver})\circ\sigma = 0, \\ (\nabla^S_{X^{hor}}Y^{hor})\circ\sigma = (\nabla_X Y)^{hor}\circ\sigma + \frac{1}{2}(R(X,Y)\sigma)^{ver}\circ\sigma, \\ (\nabla^S_{\eta^{ver}}Y^{hor})\circ\sigma = \frac{1}{2}\sum_{i=1}^n g(R(E_i,Y)\sigma,\eta)(E_i^{hor}\circ\sigma), \\ (\nabla^S_{Y^{hor}}\eta^{ver})\circ\sigma = \frac{1}{2}\sum_{i=1}^n g(R(E_i,Y)\sigma,\eta)(E_i^{hor}\circ\sigma) + (\nabla_Y\eta)^{ver}\circ\sigma, \end{cases} \tag{2.2}$$

where $\{E_i\}_{i=1}^n$ is a local $g$-orthonormal frame.

Since any $\sigma\in\Gamma^\infty(P)$ determines an embedded submanifold $\sigma : M\to(P,g^S)$, we now consider the second fundamental form of the submanifold $\sigma(M)$ to study its extrinsic geometry and also the geometry of $M$ equipped with the induced metric $\tilde{g}=\sigma^*g^S$.

For $X, Y$ vector fields in $M$ the orthogonal decomposition into the components tangent and normal to the submanifold $\sigma(M)$ of $\nabla^S_X(\sigma_*\circ Y)$ gives rise to the Levi Civita connection of the induced metric $\tilde{g}$ and to the second fundamental form

$$\alpha_\sigma : \Gamma^\infty(TM)\times\Gamma^\infty(TM)\to\Gamma^\infty(\sigma^*TP)$$

as usual

$$\nabla^S_X(\sigma_*\circ Y) = \sigma_*\circ\tilde{\nabla}_X Y + \alpha_\sigma(X,Y)$$

and the mean curvature vector field is then the vector field on $P$ along $\sigma$ defined by

$$\tau(\sigma) = \sum_{i=1}^n \alpha_\sigma(\tilde{E}_i,\tilde{E}_i),$$

where $\{\tilde{E}_i\}_{i=1}^n$ is a local $\tilde{g}$-orthonormal frame.

By the definition of the vertical and horizontal lifts we obtain that

$$\sigma_* \circ Y = Y^{hor} \circ \sigma + (\nabla_Y \sigma)^{ver} \circ \sigma \tag{2.3}$$

and therefore, using (2.2)

$$\begin{aligned}\nabla^S_X(\sigma_* \circ Y) = \; & (\nabla_X Y)^{hor} \circ \sigma + \frac{1}{2}(R(X, Y)\sigma)^{ver} \circ \sigma \\ & + \frac{1}{2}\sum_{i=1}^{n} g(R(E_i, Y)\sigma, \nabla_X \sigma)(E_i^{hor} \circ \sigma) \\ & + \frac{1}{2}\sum_{i=1}^{n} g(R(E_i, X)\sigma, \nabla_Y \sigma)(E_i^{hor} \circ \sigma) \\ & + (\nabla_X \nabla_Y \sigma)^{ver} \circ \sigma. \end{aligned} \tag{2.4}$$

By (2.3) a vector $\xi \in T_\sigma P$ is tangent to the submanifold $\sigma(M)$ if and only if $\xi = X^{hor} + \eta^{ver}$ with $\eta = \nabla_X \sigma$ and by the definition of the Sasaki metric $\xi = X^{hor} + \eta^{ver}$ is normal to the submanifold if and only if for all vector field $Y$ on $M$

$$g(Y, X) + g(\nabla_Y \sigma, \eta) = 0.$$

Consequently, a vector $\xi = X^{hor} + \eta^{ver}$ tangent to the submanifold $\sigma(M)$ is completely determined by $X$ and if $\xi$ is normal to $\sigma(M)$ then it is completely determined by $\eta$. Summing up we have shown the following

**Proposition 2.1** *Given a section $\sigma \in \Gamma^\infty(P)$ of a tensor bundle $\pi : P \to M$ over a Riemannian manifold $(M, g)$ the second fundamental form and the mean curvature vector field of the corresponding embedded submanifold $\sigma(M) \subset (P, g^S)$ are given, respectively, by*

$$\begin{aligned}\alpha_\sigma(X, Y) = \; & (\nabla_X Y - \tilde{\nabla}_X Y)^{hor} \circ \sigma \\ & + \frac{1}{2}\sum_{i=1}^{n} g(R(E_i, Y)\sigma, \nabla_X \sigma)(E_i^{hor} \circ \sigma) \\ & + \frac{1}{2}\sum_{i=1}^{n} g(R(E_i, X)\sigma, \nabla_Y \sigma)(E_i^{hor} \circ \sigma) \\ & + \frac{1}{2}(R(X, Y)\sigma)^{ver} \circ \sigma + (\nabla_X \nabla_Y \sigma - \nabla_{\tilde{\nabla}_X Y}\sigma)^{ver} \circ \sigma \end{aligned} \tag{2.5}$$

*and*

$$\tau(\sigma) = (X_\sigma)^{hor} \circ \sigma + (\eta_\sigma)^{ver} \circ \sigma,$$

*with*

$$X_\sigma = \sum_{j=1}^{n}(\nabla_{\tilde{E}_j}\tilde{E}_j - \tilde{\nabla}_{\tilde{E}_j}\tilde{E}_j) + \sum_{i,j=1}^{n} g(R(E_i, \tilde{E}_j)\sigma, \nabla_{\tilde{E}_j}\sigma)E_i$$

*and*

$$\eta_\sigma = \sum_{i=1}^{n}\left(\nabla_{\tilde{E}_i}\nabla_{\tilde{E}_i}\sigma - \nabla_{(\tilde{\nabla}_{\tilde{E}_i}\tilde{E}_i)}\sigma\right), \tag{2.6}$$

*where* $\tilde{g} = \sigma^* g^S$ *is the metric on* $M$ *induced by the immersion and* $\{E_i\}_{i=1}^n$ *and* $\{\tilde{E}_i\}_{i=1}^n$ *are local orthonormal frames with respect to* $g$ *and* $\tilde{g}$.

If for $r > 0$ we consider the sphere subbundles $S^r P$ with fibre

$$(S^r P)_x = \{\sigma_x \in P_x \quad ; \quad \|\sigma_x\| = r\},$$

it is easy to see that the unit normal vector field of the submanifold $S^r P \subset P$ at an element $\sigma \in S^r P$ is

$$\mathcal{N}^r(\sigma) = \frac{1}{r}\sigma^{ver}(\sigma).$$

These sphere subbundles $S^r P$, with the metric induced by $g^S$, have different geometrical properties for different $r$. as can be seen for example, for the case of the tangent bundle, in the papers [9] by A. A. Borisenko and A. L. Yampol'skii and [73] by O. Kowalski and M. Sekizawa. For this reason, we are going to consider sections of the general sphere subbundles $S^r P$ and not only unit sections.

The second fundamental form $\alpha^r$ of the submanifold $\sigma : M \to (S^r P, g^S)$ is given by

$$\nabla^r_X(\sigma_* \circ Y) = \sigma_* \circ \tilde{\nabla}_X Y + \alpha^r_\sigma(X, Y)$$

and the mean curvature vector field is then

$$\tau^r(\sigma) = \sum_{i=1}^{n} \alpha^r_\sigma(\tilde{E}_i, \tilde{E}_i),$$

where

$$\nabla^r_X(\sigma_* \circ Y) = \nabla^S_X(\sigma_* \circ Y) - g(\nabla^S_X(\sigma_* \circ Y), \mathcal{N}^r \circ \sigma)(\mathcal{N}^r \circ \sigma).$$

Using now (2.4) and the property $g(\nabla_X \sigma, \sigma) = 0$, the covariant derivative of the Sasaki metric on the sphere subbundle can be written as

$$\nabla^r_X(\sigma_* \circ Y) = \nabla^S_X(\sigma_* \circ Y) + \frac{1}{r^2} g(\nabla_Y \sigma, \nabla_X \sigma)\sigma^{ver} \circ \sigma.$$

So, we have shown that

**Proposition 2.2** *With the same notation as in Proposition 2.1, if* $\sigma \in \Gamma^\infty(S^r P)$ *the second fundamental form of the embedded submanifold* $\sigma(M) \subset (S^r P, g^S)$ *is*

$$\alpha^r_\sigma(X, Y) = \alpha_\sigma(X, Y) + \frac{1}{r^2} g(\nabla_Y \sigma, \nabla_X \sigma)\sigma^{ver} \circ \sigma \tag{2.7}$$

*and the mean curvature vector field*

$$\tau^r(\sigma) = (X_\sigma)^{hor} \circ \sigma + (\eta^r_\sigma)^{ver} \circ \sigma,$$

*with*

$$\eta^r_\sigma = \eta_\sigma + \frac{1}{r^2} \sum_{i=1}^{n} g(\nabla_{\tilde{E}_i} \sigma, \nabla_{\tilde{E}_i} \sigma)\sigma. \tag{2.8}$$

**Notation** For $\sigma \in \Gamma^\infty(S^r P)$ we will denote by $\mathcal{D}_\sigma$ the set of sections of the one-dimensional subbundle of $P$ determined by $\sigma$.

**Corollary 2.1** *Given a section* $\sigma \in \Gamma^\infty(P)$ *of a tensor bundle* $\pi : P \to M$ *over a Riemannian manifold* $(M, g)$ *the corresponding embedded submanifold* $\sigma(M) \subset (P, g^S)$ *is totally geodesic if and only if for every vector field X of M*

$$\nabla_X \nabla_X \sigma - \nabla_{\tilde{\nabla}_X X} \sigma = 0 \tag{2.9}$$

*and* $\sigma(M)$ *is minimal if and only if*

$$\eta_\sigma = \sum_{i=1}^{n} \left( \nabla_{\tilde{E}_i} \nabla_{\tilde{E}_i} \sigma - \nabla_{(\tilde{\nabla}_{\tilde{E}_i} \tilde{E}_i)} \sigma \right) = 0. \tag{2.10}$$

*If* $\sigma$ *is a section of the sphere subbundle* $S^r P$ *then the submanifold* $\sigma(M) \subset (S^r P, g^S)$ *is totally geodesic if and only if for every vector field X of M*

$$\nabla_X \nabla_X \sigma - \nabla_{\tilde{\nabla}_X X} \sigma \in \mathcal{D}_\sigma$$

*and* $\sigma(M)$ *is minimal if and only if* $\eta_\sigma \in \mathcal{D}_\sigma$.

***Proof*** The submanifold is totally geodesic if and only if the second fundamental form vanishes and, since $\alpha_\sigma(X, Y)$ is normal to $\sigma(M)$, this is equivalent to

$$\frac{1}{2}R(X, Y)\sigma + \nabla_X \nabla_Y \sigma - \nabla_{\tilde{\nabla}_X Y}\sigma = 0,$$

for all $X, Y$. The equivalence with (2.9) is a consequence of the symmetry of $\alpha_\sigma$. The submanifold is minimal if and only if the mean curvature vector field vanishes for which again the vanishing of its vertical part is an equivalent condition.

The corresponding equalities for a section of the sphere bundle are obtained by using the expressions of $\alpha^r$ and $\tau^r$ instead of those of $\alpha$ and $\tau$ and having into account that the vertical part of $\alpha_\sigma(X, X)$ is in $\mathscr{D}_\sigma$ if and only if

$$\nabla_X \nabla_X \sigma - \nabla_{\tilde{\nabla}_X X}\sigma = \frac{1}{r^2} g(\nabla_X \nabla_X \sigma - \nabla_{\tilde{\nabla}_X X}\sigma, \sigma)\sigma,$$

and since $g(\nabla_Z \sigma, \sigma) = 0$ for all vector field $Z$, the condition is equivalent to

$$\nabla_X \nabla_X \sigma - \nabla_{\tilde{\nabla}_X X}\sigma = -\frac{1}{r^2} g(\nabla_X \sigma, \nabla_X \sigma)\sigma. \tag{2.11}$$

Analogously $\eta_\sigma \in \mathscr{D}_\sigma$ if and only if

$$\eta_\sigma = -\frac{1}{r^2}\sum_{i=1}^{n} g(\nabla_{\tilde{E}_i}\sigma, \nabla_{\tilde{E}_i}\sigma, )\sigma. \tag{2.12}$$

□

## 2.2 Minimal Sections of Tensor Bundles and Sphere Subbundles

In the previous results, the second fundamental form is expressed in terms of the Levi Civita connections of both metrics $g$ and $\tilde{g} = \sigma^* g^S$ and then the dependence on the section $\sigma$ is partially concealed. It will be useful to write $\alpha_\sigma$ and $\eta_\sigma$ in terms of the Levi Civita connection of $g$.

**Definition 2.1** For a tensor bundle $\pi : P \to M$ over a Riemannian manifold we can define the following differential operators:

1. The covariant derivative acting on sections of the bundle is the differential operator $\nabla : \Gamma^\infty(P) \to \Gamma^\infty(P \otimes T^*M)$ given by $(\nabla\sigma)(X) = \nabla_X \sigma$ for $\sigma \in \Gamma^\infty(P)$ and $X \in \Gamma^\infty(TM)$.
2. The divergence acting on sections of $P \otimes T^*M$ is the differential operator $\nabla^* : \Gamma^\infty(P \otimes T^*M) \to \Gamma^\infty(P)$ obtained by a tensor contraction of the covariant

derivative. For $K \in \Gamma^\infty(P \otimes T^*M)$ and $\{E_i\}_{i=1}^n$ an orthonormal local frame, the local expression of $\nabla^* K$ is

$$\nabla^* K = -\sum_{i=1}^{n} (\nabla_{E_i} K) E_i = -\sum_{i=1}^{n} \Big\{ \nabla_{E_i} K(E_i) - K(\nabla_{E_i} E_i) \Big\}.$$

3. The rough Laplacian is defined as the composition $\nabla^*\nabla : \Gamma^\infty(P) \to \Gamma^\infty(P)$.

**Proposition 2.3** *If $K \in \Gamma^\infty(P \otimes T^*M)$ and $\sigma \in \Gamma^\infty(P)$ then*

$$g(\nabla^* K, \sigma) = g(K, \nabla \sigma) + \nabla^*(K[\sigma]) \tag{2.13}$$

*where $K[\sigma] \in \Gamma^\infty(T^*M)$ is defined by $K[\sigma](X) = g(K(X), \sigma)$ for all $X \in \Gamma^\infty(TM)$.*

***Proof*** Let $\{E_i\}_{i=1}^n$ be an orthonormal frame defined in an open subset of $M$, the divergence of the 1-form $K[\sigma]$ in this open subset is given by

$$\begin{aligned}
\nabla^*(K[\sigma]) &= -\sum_{i=1}^{n} \Big\{ \nabla_{E_i} K[\sigma](E_i) - K[\sigma](\nabla_{E_i} E_i) \Big\} \\
&= -\sum_{i=1}^{n} \Big\{ \nabla_{E_i} g(K(E_i), \sigma) - g(K(\nabla_{E_i} E_i), \sigma) \Big\} \\
&= -\sum_{i=1}^{n} \Big\{ g(\nabla_{E_i} K(E_i), \sigma) + g(K(E_i), \nabla_{E_i} \sigma) - g(K(\nabla_{E_i} E_i), \sigma) \Big\}.
\end{aligned}$$

To conclude we only need to take into account that

$$g(K, \nabla \sigma) = \sum_{i=1}^{n} g(K(E_i), \nabla_{E_i} \sigma)$$

and the definition of $\nabla^* K$. □

The following Lemma gives the relationship between the covariant derivatives of two metrics.

**Lemma 2.1** *Let $g$ and $\tilde{g}$ be two Riemannian metrics on a manifold $M$ and let $L$ be the endomorphism field defined by $\tilde{g}(X, Y) = g(L(X), Y)$ for all $X, Y \in \Gamma^\infty(TM)$. The corresponding Levi Civita connections are related by the expression*

$$\tilde{\nabla}_X Y = \nabla_X Y + \frac{1}{2} L^{-1}\big((\nabla_X L)(Y) + (\nabla_Y L)(X) - (\beta L)(X, Y)\big),$$

*where* $(\beta L)(X, Y)$ *is the vector field defined by*

$$g((\beta L)(X, Y), Z) = g(X, (\nabla_Z L)(Y))$$

*for all* $Z \in \Gamma^\infty(TM)$.

***Proof*** The Koszul formula applied to the metric $\tilde{g}$ reads

$$2\tilde{g}(\tilde{\nabla}_X Y, Z) = X\tilde{g}(Y, Z) + Y\tilde{g}(X, Z) - Z\tilde{g}(X, Y) \\ +\tilde{g}([X, Y], Z) + \tilde{g}([Z, X], Y) + \tilde{g}(X, [Z, Y])$$

and by definition of $L$ it is equivalent to

$$2g(L(\tilde{\nabla}_X Y), Z) = Xg(L(Y), Z) + Yg(L(X), Z) - Zg(L(X), Y) \\ +g(L([X, Y]), Z) + g(L([Z, X]), Y) + g(L(X), [Z, Y]).$$

We can develop the first three terms of the sum to obtain

$$Xg(L(Y), Z) = g((\nabla_X L)(Y), Z) + g(L(\nabla_X Y), Z) + g(LY, \nabla_X Z), \\ Yg(L(X), Z) = g((\nabla_Y L)(X), Z) + g(L(\nabla_Y X), Z) + g(LX, \nabla_Y Z), \\ -Zg(L(X), Y) = -g((\nabla_Z L)(X), Y) - g(L(\nabla_Z X), Y) - g(LX, \nabla_Z Y).$$

Taking these expressions into account, it is easy to check that

$$2g(L(\tilde{\nabla}_X Y), Z) = 2g(L(\nabla_X Y), Z) + g((\nabla_X L)(Y), Z) + g((\nabla_Y L)(X), Z) \\ -g((\nabla_Z L)(X), Y)$$

from where the result holds. □

**Theorem 2.1** *Given a section* $\sigma \in \Gamma^\infty(P)$ *of a tensor bundle* $\pi : P \to M$ *over a Riemannian manifold* $(M, g)$, *let us denote by* $K_\sigma$ *the section of* $P \otimes T^*M$ *acting on vector fields as*

$$K_\sigma(Y) = f_\sigma \nabla_{L_\sigma^{-1}(Y)}\sigma,$$

*where we represent by* $L_\sigma$ *the endomorphism field defined by*

$$g(L_\sigma(X), Y) = \sigma^* g^S(X, Y) = g(X, Y) + g(\nabla_X \sigma, \nabla_Y \sigma) \tag{2.14}$$

*and* $f_\sigma = \sqrt{\det L_\sigma}$. *Then*

$$\eta_\sigma = -\frac{1}{f_\sigma}\nabla^* K_\sigma.$$

***Proof*** By the definition of $\eta_\sigma$ in (2.6), we need to calculate terms of the form $\nabla_X\nabla_X\sigma - \nabla_{\tilde{\nabla}_X X}\sigma$ for the metric $\tilde{g} = \sigma^* g^S$, which are related with $K_\sigma$ as follows

$$\begin{aligned}(\nabla_X K_\sigma)(L_\sigma(X)) &= \nabla_X K_\sigma(L_\sigma(X)) - K_\sigma(\nabla_X L_\sigma(X)) \\ &= \nabla_X(f_\sigma \nabla_X\sigma) - f_\sigma \nabla_{L_\sigma^{-1}(\nabla_X L_\sigma(X))}\sigma \\ &= f_\sigma(\nabla_X\nabla_X\sigma - \nabla_{\tilde{\nabla}_X X}\sigma) + \nabla_{Q_\sigma(X,X)}\sigma, \end{aligned} \tag{2.15}$$

where

$$\begin{aligned}Q_\sigma(X,X) &= X(f_\sigma)X - \frac{1}{2}f_\sigma(L_\sigma)^{-1}((\beta L_\sigma)(X,X)) \\ &= \frac{1}{2}f_\sigma\Big(\mathrm{tr}((L_\sigma)^{-1}\circ\nabla_X L_\sigma)X - (L_\sigma)^{-1}((\beta L_\sigma)(X,X))\Big).\end{aligned}$$

For the last equality in (2.15) we have used that, by previous Lemma,

$$\tilde{\nabla}_X X = \nabla_X X + (L_\sigma)^{-1}\Big((\nabla_X L_\sigma)(X) - \frac{1}{2}(\beta L_\sigma)(X,X)\Big).$$

First we show by straightforward computations that the trace of $Q_\sigma$ vanishes.

$$\begin{aligned}\frac{2}{f_\sigma}\sum_{j=1}^{n} Q_\sigma(\tilde{E}_j,\tilde{E}_j) &= \sum_{j=1}^{n}\sum_{i=1}^{n} g((\nabla_{\tilde{E}_j}L_\sigma)E_i, L_\sigma^{-1}(E_i))\tilde{E}_j \\ &\quad - \sum_{j=1}^{n}\sum_{i=1}^{n} g((\nabla_{E_i}L_\sigma)\tilde{E}_j,\tilde{E}_j)L_\sigma^{-1}(E_i) \\ &= \sum_{i,j,k=1}^{n} g((\nabla_{\tilde{E}_j}L_\sigma)E_i,\tilde{E}_k)g(\tilde{E}_k,E_i)\tilde{E}_j \\ &\quad - \sum_{i,j,k=1}^{n} g((\nabla_{E_i}L_\sigma)\tilde{E}_j,\tilde{E}_j)g(\tilde{E}_k,E_i)\tilde{E}_k = 0,\end{aligned} \tag{2.16}$$

where for the second equality we have used that

$$L_\sigma^{-1}(E_i) = \sum_{k=1}^{n}\tilde{g}(L_\sigma^{-1}(E_i),\tilde{E}_k)\tilde{E}_k = \sum_{k=1}^{n} g(\tilde{E}_k,E_i)\tilde{E}_k.$$

Combining (2.6), (2.15) and (2.16) we have

$$\eta_\sigma = \frac{1}{f_\sigma}\sum_{j=1}^{n}(\nabla_{\tilde{E}_j}K_\sigma)(L_\sigma(\tilde{E}_j)).$$

But

$$L_\sigma(\tilde{E}_j) = \sum_{i=1}^{n} g(L_\sigma(\tilde{E}_j), E_i)E_i = \sum_{i=1}^{n} \tilde{g}(\tilde{E}_j, E_i)E_i$$

and therefore

$$\eta_\sigma = \frac{1}{f_\sigma}\sum_{i=1}^{n}\sum_{j=1}^{n} \tilde{g}(\tilde{E}_j, E_i)(\nabla_{\tilde{E}_j} K_\sigma)E_i = \frac{1}{f_\sigma}\sum_{i=1}^{n}(\nabla_{E_i} K_\sigma)E_i,$$

as stated. □

As a consequence of the Theorem above, the condition for the immersion $\sigma : M \to (P, g^S)$ to be minimal can be written as a second order differential equation in its covariant derivate $\nabla\sigma$.

**Corollary 2.2** *The submanifold $\sigma(M) \subset (P, g^S)$, for $\sigma \in \Gamma^\infty(P)$, is minimal if and only if*

$$\nabla^* K_\sigma = 0. \tag{2.17}$$

*For $\sigma \in \Gamma^\infty(S^r P)$, the submanifold $\sigma(M) \subset (S^r P, g^S)$ is minimal if and only if $\nabla^* K_\sigma \in \mathcal{D}_\sigma$ which is equivalent to*

$$\nabla^* K_\sigma = \frac{1}{r^2} g(K_\sigma, \nabla\sigma)\sigma. \tag{2.18}$$

***Proof*** It follows from Corollary 2.1 and Theorem 2.1. For the last assertion, if $\|\sigma\| = r$, the condition $\nabla^* K_\sigma \in \mathcal{D}_\sigma$ implies

$$\nabla^* K_\sigma = \frac{1}{r^2} g(\nabla^* K_\sigma, \sigma)\sigma.$$

But, by (2.13)

$$g(\nabla^* K_\sigma, \sigma) = g(K_\sigma, \nabla\sigma) + \nabla^*(K_\sigma[\sigma])$$

and $K_\sigma[\sigma] = 0$ since $\sigma$ is of constant length. □

**Definition 2.2** Let $\pi : P \to M$ be a tensor bundle over a Riemannian manifold. A section $\sigma$ will be said minimal if $\sigma(M) \subset (P, g^S)$ is a minimal submanifold. A section of constant length $r$, with $r > 0$, will be said $r$-minimal if $\sigma(M) \subset (S^r P, g^S)$ is a minimal submanifold.

The definition of $r$-minimal section above extends the concept of minimal unit section studied in the previous literature on the volume of sections.

The Riemannian version of the classical Divergence Theorem and Green Theorem can be seen for instance in [86, p. 71]. Let's recall that the divergence of a vector field $X$ is defined as the function $\operatorname{div}X = -\operatorname{tr}(\nabla X)$.

**Theorem 2.2 (Divergence Theorem: Green Theorem)** *Let $X$ be a vector field with compact support on a Riemannian manifold $(M, g)$ then*

$$\int_M \operatorname{div}X \, \mathrm{dv}_g = 0.$$

*Assume that $(M, g)$ is a manifold with boundary $\partial M \neq \emptyset$ and that we represent by $\overline{g}$ the Riemannian metric on $\partial M$ induced by $g$. If $N$ is the outwards unit normal on $\partial M$, then*

$$\int_M \operatorname{div}X \, \mathrm{dv}_g = \int_{\partial M} g(X, N) \, \mathrm{dv}_{\overline{g}}.$$

The theorem above has an immediate translation in terms of 1-forms, which we will also use. We only need to take into account that if $\omega$ is a 1-form and $X$ is the corresponding vector field, i. e. $\omega(Y) = g(X, Y)$, for all $Y$, then $\operatorname{div} X = \nabla^*\omega$.

**Proposition 2.4** *On a closed manifold a section of a tensor bundle is minimal if and only if it is parallel. The same conclusion is also true for sections with compact support if the manifold is not compact.*

***Proof*** By definition, $\nabla\sigma = 0$ implies that $K_\sigma = 0$ and consequently a parallel section $\sigma$ is minimal by the Corollary above. Let's assume reciprocally that $\sigma$ is minimal, which is equivalent to $\nabla^* K_\sigma = 0$, then by Proposition 2.13 we have

$$g(K_\sigma, \nabla\sigma) + \nabla^*(K_\sigma[\sigma]) = 0.$$

Applying the Divergence Theorem to the 1-form $K_\sigma[\sigma]$ we obtain for $M$ closed (or for $\sigma$ with compact support, otherwise) that

$$\int_M g(K_\sigma, \nabla\sigma)\mathrm{dv}_g = 0.$$

The endomorphism field $L_\sigma^{-1}$ defined in Theorem 2.1 is symmetric and positive definite and then its lowest eigenvalue $\lambda$ is positive. Since $g(K_\sigma, \nabla\sigma) \geq \lambda f_\sigma g(\nabla\sigma, \nabla\sigma) \geq 0$ and $\lambda > 0$, $f_\sigma > 0$ then $\nabla\sigma = 0$. □

*Remark* In accordance with the Green Theorem, when the manifold $M$ is compact with boundary $\partial M \neq \emptyset$ and if $N$ is the vector field normal to $\partial M$, to obtain the conclusion $\nabla\sigma = 0$ for the solutions of the differential equation (2.17) we need to assume that on $\partial M$ the section fulfils either the Dirichlet condition $\sigma = 0$ or the Neumann type condition $K_\sigma(N) = 0$.

## 2.3 First Variation of the Volume of Vector Fields: Minimal Vector Fields

For a smooth vector field $V \in \Gamma^\infty(TM)$ on a Riemannian manifold $(M, g)$ the volume of $V(M) \subset (TM, g^S)$ is equal to the volume of the Riemannian manifold $(M, V^*g^S)$ and can be expressed in terms of the endomorphism field $L_V$ relating the two metrics $g$ and $V^*g^S$ by $V^*g^S(X, Y) = g(L_V(X), Y)$. Accordingly with (2.14),

$$V^*g^S(X, Y) = g(X, Y) + g(\nabla_X V, \nabla_Y V) = g(X, Y) + g(\nabla V(X), \nabla V(Y)),$$

where the covariant derivative operator in Definition 2.1 in this particular case sends any vector field $V$ to the endomorphism field $\nabla V$ given by $\nabla V(X) = \nabla_X V$. Then

$$L_V = \mathrm{Id} + (\nabla V)^t \circ \nabla V \tag{2.19}$$

and we denote the square root of its determinant by

$$f_V = \sqrt{\det L_V}. \tag{2.20}$$

**Definition 2.3** The volume of a vector field $V$ on a Riemannian manifold $(M, g)$ is the volume of the submanifold $V(M)$ of $(TM, g^S)$. We can define the volume functional $\mathrm{Vol} : \Gamma^\infty(TM) \to \mathbf{R} \cup \{+\infty\}$ that is given by

$$\mathrm{Vol}(V) = \int_M f_V \mathrm{dv}_g.$$

As a direct consequence of the expression of $V^*g^S$, the Definition 2.3 and the Proposition 2.2 we have the following Lemma

**Lemma 2.2** *Let $(M, g)$ be a Riemannian manifold and for $\lambda \in \mathbf{R}$, $\lambda \neq 0$, let us consider the homothetic metric $\lambda^2 g$.*

1. *For every vector field, $V^*(\lambda^2 g)^S = \lambda^2 V^* g^S$.*
2. *If $Vol^\lambda$ represents the volume functional corresponding to the metric $\lambda^2 g$ and $n$ is the dimension of $M$ then $\mathrm{Vol}^\lambda(V) = \lambda^n \mathrm{Vol}(V)$, for all $V$.*
3. *$V$ is a minimal vector field on $(M, g)$ if and only if $V$ is a minimal vector field on $(M, \lambda^2 g)$.*
4. *If we assume that $V$ has length $r$ in the metric $g$ then, $V$ is $r$-minimal as a vector field on $(M, g)$ if and only if it is $\lambda r$-minimal as a vector field on $(M, \lambda^2 g)$.*

*Remark* It's clear from the definition that for a Riemannian manifold $(M, g)$ of finite volume the volume functional is bounded below by the volume of $(M, g)$ and that this bound is attained only by parallel vector fields .

Let's compute the first variation of the functional Vol, at a vector field $V$ in the direction of a variational vector field $A$, for which we will use the functorial notation $(T_V\mathrm{Vol})(A)$, tangent map of the map Vol at $V$ acting on $A \in T_V(\Gamma^\infty(TM)) = \Gamma^\infty(TM)$.

In the sequel we will use the notation $\langle L, K\rangle$ for the pointwisse inner product induced by $g$ on the space of endomorphism fields, i. e. $\langle L, K\rangle = \mathrm{tr}(K^t \circ L)$.

**Proposition 2.5** *Let $V \in \Gamma^\infty(TM)$ be a smooth vector field and let $A \in T_V(\Gamma^\infty(TM))$ be a tangent vector at $V$. The first variation of the functional* Vol *at $V$, in the direction of the variational vector field $A$ is given by*

$$(T_V\mathrm{Vol})(A) = \int_M (T_V f)(A)\mathrm{dv}_g = \int_M g(\nabla^* K_V, A)\mathrm{dv}_g - \int_M \nabla^*(K_V[A])\mathrm{dv}_g,$$

*where $K_V$ is the endomorphism field $K_V = f_V \nabla V \circ L_V^{-1}$.*

***Proof*** Let $V : I \to \Gamma^\infty(TM)$ be a smooth curve from some open interval $I$ containing 0, such that $V(0) = V$ and $V'(0) = A$. If we denote $L(s) = L_{V(s)}$, $f(s) = f_{V(s)}$ and $K(s) = K_{V(s)}$ then

$$f' = \frac{1}{2} f\ \mathrm{tr}(L' \circ L^{-1}), \qquad L' = (\nabla V')^t \circ \nabla V + (\nabla V)^t \circ \nabla V'$$

and therefore

$$f' = \frac{1}{2} f \big(\mathrm{tr}((\nabla V')^t \circ \nabla V \circ L^{-1}) + \mathrm{tr}((\nabla V)^t \circ \nabla V' \circ L^{-1})\big).$$

Since $L$ is symmetric,

$$\mathrm{tr}((\nabla V)^t \circ \nabla V' \circ L^{-1}) = \mathrm{tr}(L^{-1} \circ (\nabla V)^t \circ \nabla V') = \mathrm{tr}((\nabla V')^t \circ \nabla V \circ L^{-1}).$$

On the other hand,

$$\mathrm{tr}((\nabla V')^t \circ \nabla V \circ L^{-1}) = \langle \nabla V \circ L^{-1}, \nabla V'\rangle$$

and, by the definition of $K$, we get

$$f' = \langle K, \nabla V'\rangle. \tag{2.21}$$

If we apply (2.13) with $\sigma = V'$, the derivative of $f$ can be also written as

$$f' = g(\nabla^* K, V') - \nabla^*(K[V']). \tag{2.22}$$

To end the proof we only need to evaluate (2.22) at $s = 0$, to obtain

$$(T_V f)(A) = f'(0) = g(\nabla^* K_V, A) - \nabla^*(K_V[A]),$$

and to integrate over $M$. □

After computing the first variation of the volume functional we can characterise its critical points and also the critical points of its restriction to vector fields of constant length. For $r > 0$, we will denote the sphere bundle $S^r TM$ by $T^r M$. We will start with the case of compact manifolds without boundary.

**Proposition 2.6** *Let $(M, g)$ be a closed Riemannian manifold then*

*(1) A vector field is a critical point of the volume functional if and only if $V(M)$ is a minimal submanifold of $(TM, g^S)$.*
*(2) A vector field $V$ of constant length $r$ is a critical point of the volume functional restricted to $\Gamma^\infty(T^r M)$ if and only if $V(M)$ is a minimal submanifold of $(T^r M, g^S)$. Or in other words, if and only if $V$ is $r$-minimal in the sense of Definition 2.2.*

***Proof*** For a closed manifold, using the Divergence Theorem, Proposition 2.5 implies that $V$ is a critical point if and only if

$$\int_M g(\nabla^* K_V, A) \mathrm{d}v_g = 0 \tag{2.23}$$

for all vector field $A$ which is equivalent to $\nabla^* K_V = 0$. Then, part 1 follows from (2.17).

If $\|V\| = r$, it satisfies the weaker condition of being a critical point for variations by vector fields in $\Gamma^\infty(T^r M)$ if and only if equality (2.23) holds for all $A \in T_V(\Gamma^\infty(T^r M)) = \mathscr{D}_V^\perp$ which is equivalent to the vanishing of the projection of $\nabla^* K_V$ onto $\mathscr{D}_V^\perp$, which is in turn equivalent to $\nabla^* K_V \in \mathscr{D}_V$. Part 2 follows from Corollary 2.2. As in (2.18), the condition can also be written in the form

$$\nabla^* K_V = \frac{1}{r^2} \langle K_V, \nabla V \rangle V. \tag{2.24}$$

□

As a direct consequence of part 1 of Proposition 2.6 and Proposition 2.4 we have the following

**Corollary 2.3** *On a closed manifold the only critical points of the volume functional* Vol : $\Gamma^\infty(TM) \to \mathbf{R}$ *are parallel vector fields.*

*Remark* From Proposition 2.6, if a vector field is a critical point of the volume functional for vector fields (resp. for vector fields of length $r$) then it is also a critical point of the volume for general immersions of $M$ into $TM$ (rep. $T^r M$); this fact was shown in [47] directly without using Corollary 2.2.

As it occurs in the case of the submanifolds of a Riemannian manifold, to be minimal and to be a critical point of the volume are equivalent properties only for closed manifolds. For a general manifold, the minimality of $V$ (resp. the $r$-minimality, if $V$ has constant length $r$) is equivalent to

$$(T_V\mathrm{Vol})(A) = -\int_M \nabla^*(K_V[A])\mathrm{dv}_g$$

for all vector field $A$ (resp. for $A \in \mathscr{D}_V^{\perp}$).

By the Divergence Theorem, if $M$ is compact with nonempty boundary $\partial M$ and outwards unit normal $N$, a vector field $V$ is minimal (resp. $r$-minimal) if and only if for all $A$ (resp. for $A \in \mathscr{D}_V^{\perp}$)

$$(T_V\mathrm{Vol})(A) = \int_{\partial M} g(A, K_V(N))\mathrm{dv}_{\overline{g}}.$$

Then we have the following characterisation

**Proposition 2.7** *On a compact manifold with boundary a vector field $V$ with $K_V(N) = 0$ is minimal (resp. the $r$-minimal, if $V$ has constant length $r$) if and only if it is a critical point of the volume functional. The same result is true on open manifolds for vector fields $V$ such that $K_V$ has compact support.*

Without additional conditions on $K_V$, what we get is that minimality is equivalent to the vanishing of the derivative of the volume functional for certain variations.

**Proposition 2.8** *On a compact manifold with boundary minimal vector fields (resp. $r$-minimal) are critical points of the volume functional restricted to variations whose variational vector field vanishes at the boundary; in particular for variations that let the boundary fixed.*

*On an open manifold, minimal vector fields (resp. $r$-minimal) are critical points of the volume functional restricted to variations whith variational vector field of compact support; in particular for variations with compact support.*

## 2.4 Second Variation of the Volume of Vector Fields

In this section we are going to assume that the manifold is closed and we are going to compute the expression of the Hessian of the Volume functional with the aim of studying later on the stability of the critical points. The stability of minimal vector fields is guaranteed by Corollary 2.3 since parallel vector fields are the absolute minimisers of the volume, hence we are going to compute the Hessian at each $r$-minimal vector field. The proof follows the same lines that the one in our joint paper with Llinares-Fuster [46].

**Theorem 2.3** *Let $M$ be a closed manifold. The Hessian of the Volume functional at an $r$-minimal vector field $V$ acting on $A \in \mathscr{D}_V^{\perp}$ is given by*

$$
\begin{aligned}
(Hess\mathrm{Vol})_V(A) = &-\frac{1}{r^2}\int_M \|A\|^2 \langle K_V, \nabla V\rangle \mathrm{dv}_g \\
&+\int_M \frac{1}{f_V}\Big(\langle K_V, \nabla A\rangle^2 - \langle K_V \circ (\nabla A)^t \circ K_V, \nabla A\rangle\Big)\mathrm{dv}_g \\
&-\int_M \langle K_V \circ (\nabla V)^t \circ \nabla A \circ L_V^{-1}, \nabla A\rangle \mathrm{dv}_g \\
&+\int_M f_V \langle \nabla A \circ L_V^{-1}, \nabla A\rangle \mathrm{dv}_g. \qquad (2.25)
\end{aligned}
$$

***Proof*** Let $V : I \to \Gamma^{\infty}(T^r M)$ be a smooth curve from some open interval $I$ containing 0, such that $V(0) = V$ and $V'(0) = A$. With the same notation as in Proposition 2.5 we have that

$$(Hess\mathrm{Vol})_V(A) = \int_M f''(0)\mathrm{dv}_g.$$

Since by (2.21) $f' = \langle K, \nabla V'\rangle$, the second derivative of $f$ at $s = 0$ is

$$f''(0) = \langle K(0), \nabla V''(0)\rangle + \langle K'(0), \nabla V'(0)\rangle.$$

To compute $K'(0)$ we take into account that

$$(L^{-1})'(0) = -L_V^{-1} \circ (\nabla A)^t \circ \nabla V \circ L_V^{-1} - L_V^{-1} \circ (\nabla V)^t \circ \nabla A \circ L_V^{-1}$$

and then

$$
\begin{aligned}
K'(0) = &\langle K_V, \nabla A\rangle \nabla V \circ L_V^{-1} + f_V \nabla A \circ L_V^{-1} - f_V \nabla V \circ L_V^{-1} \circ (\nabla A)^t \circ \nabla V \circ L_V^{-1} \\
&- f_V \nabla V \circ L_V^{-1} \circ (\nabla V)^t \circ \nabla A \circ L_V^{-1}. \qquad (2.26)
\end{aligned}
$$

Consequently

$$
\begin{aligned}
f''(0) = &\langle K_V, \nabla V''(0)\rangle + \langle K_V, \nabla A\rangle\langle \nabla V \circ L_V^{-1}, \nabla A\rangle + f_V\langle \nabla A \circ L_V^{-1}, \nabla A\rangle \\
&- f_V\langle \nabla V \circ L_V^{-1} \circ (\nabla A)^t \circ \nabla V \circ L_V^{-1}, \nabla A\rangle \\
&- f_V\langle \nabla V \circ L_V^{-1} \circ (\nabla V)^t \circ \nabla A \circ L_V^{-1}, \nabla A\rangle. \qquad (2.27)
\end{aligned}
$$

We only need to show that the integral of $\langle K_V, \nabla V''(0)\rangle$ is equal to the first term of (2.25) and, in particular, that it only depends of $V'(0)$. To do so, we apply (2.13) with $\sigma = V''(0)$ to obtain

$$\langle K_V, \nabla V''(0)\rangle = g(\nabla^* K_V, V''(0)) - \nabla^*(K_V[V''(0)]).$$

Since $V$ is $r$-minimal and the variation verifies $\|V(s)\| = r$, we get

$$\begin{aligned}\langle K_V, \nabla V''(0)\rangle + \nabla^*(K_V[V''(0)]) &= \frac{1}{r^2}\langle K_V, \nabla V\rangle g(V, V''(0)) \\ &= -\frac{1}{r^2}\|A\|^2\langle K_V, \nabla V\rangle. \qquad (2.28)\end{aligned}$$

Now the result follows by integration of (2.28), using Proposition 2.2 because $M$ is closed. □

**Definition 2.4** An $r$-minimal vector field $V$ is said to be stable if and only if $(Hess\mathrm{Vol})_V(A) \geq 0$ for all $A \in \mathscr{D}_V^{\perp}$.

*Remark* Although we have stated Theorem 2.25 only for closed manifolds, the expression given for the Hessian at a critical point (2.25) is also valid for compact manifolds with boundary or for open manifolds, if we assume the additional conditions on $V$ of Proposition 2.7. The reason is that under these conditions

$$\int_M \nabla^*(K_V[V''(0)])\mathrm{dv}_g = 0.$$

Without these hypotheses on $V$ we also obtain that the right hand side of (2.25) is the value of the second variation of the volume at $r$-minimal vector fields for variations fixing the boundary with variational field $A \in \mathscr{D}_V^{\perp}$, or for variations with compact support in the open case, because then $V''(0)$ vanishes on the boundary or has compact support, respectively. It's noteworthy that it's not sufficient to assume that $A$ vanishes on the boundary or that $A$ has compact support and so formally we can't write the second derivative $\int_M f''(0)\mathrm{dv}_g$ as $(Hess\mathrm{Vol})_V(A)$. Nevertheless we can define the concept of stability to refer to minimal vector fields as being stable if and only if

$$\int_M f''(0)\mathrm{dv}_g \geq 0$$

for every variation fixing the boundary or with compact support, when the manifold is compact with boundary or open, respectively.

## 2.5 The 2-Dimensional Case

In the particular case of a manifold $M$ of dimension 2, a vector field $V$ of constant length $r$ determines a co-dimension one surface $V(M)$ of $T^r M$. The explicit computations of the first and second variations can be simplified and it's possible then to obtain the following result, which can't be generalised to higher dimensional manifolds as we will see in the next chapters. It was proved for unit vector fields in [46].

**Theorem 2.4** *Let $V$ be a vector field of length $r$ on a manifold $M$ of dimension 2 then*

*(a) $V$ is $r$-minimal if and only if for every unit vector field $E$ orthogonal to $V$ defined on some open subset*

$$E\left(\frac{g(\nabla_E V, E)}{f_V}\right) = -\frac{1}{r^2} V\left(\frac{g(\nabla_V V, E)}{f_V}\right).$$

*(b) If $V$ is $r$-minimal, every vector field defined on some open subset, of length $r$ and orthogonal to $V$ is also $r$-minimal.*

*(c) Every $r$-minimal vector field is stable. Moreover, the Hessian acting on $A \in \mathscr{D}_V^{\perp}$ vanishes if and only if $A$ is of constant length.*

***Proof*** Since the dimension of $M$ is 2, $V$ is $r$-minimal if and only if $g(\nabla^* K_V, E) = 0$ for all unit vector field $E$ orthogonal to $V$ defined in some open subset of $M$. Let's compute $\nabla^* K_V$ in the local orthonormal frame $\{E, V/r\}$; the only non vanishing entries of the matrix of $\nabla V$ are

$$(\nabla V)_{11} = g(\nabla_E V, E) \quad \text{and} \quad (\nabla V)_{12} = \frac{1}{r} g(\nabla_V V, E) \tag{2.29}$$

and the matrix of $L_V$ in this frame is

$$L_V = \begin{pmatrix} 1 + g(\nabla_E V, E)^2 & \frac{1}{r} g(\nabla_V V, E) g(\nabla_E V, E) \\ \frac{1}{r} g(\nabla_V V, E) g(\nabla_E V, E) & 1 + \frac{1}{r^2} g(\nabla_V V, E)^2 \end{pmatrix} \tag{2.30}$$

from where $f_V = \sqrt{1 + g(\nabla_E V, E)^2 + \frac{1}{r^2} g(\nabla_V V, E)^2}$. The matrix of the endomorphism field $L_V^{-1}$ is

$$L_V^{-1} = \frac{1}{f_V^2} \begin{pmatrix} 1 + \frac{1}{r^2} g(\nabla_V V, E)^2 & -\frac{1}{r} g(\nabla_V V, E) g(\nabla_E V, E) \\ -\frac{1}{r} g(\nabla_V V, E) g(\nabla_E V, E) & 1 + g(\nabla_E V, E)^2 \end{pmatrix} \tag{2.31}$$

and

$$K_V = \frac{1}{f_V} \nabla V. \tag{2.32}$$

The divergence of $K_V$ verifies

$$\begin{aligned} g(\nabla^* K_V, E) &= -(\nabla_E(K_V(E)), E) - \frac{1}{r^2}(\nabla_V(K_V(V)), E) \\ &\quad + g(K_V(\nabla_E E), E) + \frac{1}{r^2} g(K_V(\nabla_V V), E) \\ &= -E\Big(\frac{g(\nabla_E V, E)}{f_V}\Big) - \frac{1}{r^2} V\Big(\frac{g(\nabla_V V, E)}{f_V}\Big) \end{aligned} \tag{2.33}$$

from where we obtain part (a). Part (b) can be easily verified by writing (2.33) with $V = rE$ and then $E = V/r$.

If $V$ is $r$-minimal and $A \in \mathscr{D}_V^{\perp}$, let's compute the integrand of the Hessian of the volume (2.25) in a local orthonormal frame of the form $\{E, V/r\}$. If we denote $a = g(A, E)$, the first term is

$$-\frac{\|A\|^2}{r^2}\langle K_V, \nabla V\rangle = -\frac{a^2}{r^2}\Big(\frac{1}{f_V} g(\nabla_E V, E)^2 + \frac{1}{r^2 f_V} g(\nabla_V V, E)^2\Big) = -\frac{a^2(f_V^2 - 1)}{r^2 f_V}. \tag{2.34}$$

The matrix of $\nabla A$ in the chosen frame is

$$\nabla A = \frac{1}{r^2}\begin{pmatrix} r^2 E(a) & r V(a) \\ ra\, g(\nabla_E E, V) & a\, g(\nabla_V E, V) \end{pmatrix} \tag{2.35}$$

and consequently, using (2.32), we get

$$(\nabla A)^t \circ K_V = \frac{1}{r^2 f_V}\begin{pmatrix} r^2 E(a) g(\nabla_E V, E) & r E(a) g(\nabla_V V, E) \\ r V(a) g(\nabla_E V, E) & V(a) g(\nabla_V V, E) \end{pmatrix}. \tag{2.36}$$

By straightforward computation using (2.36), it can be seen that the second term of the integrand of the Hessian vanishes

$$\frac{1}{f_V}\Big(\langle K_V, \nabla A\rangle^2 - \langle K_V \circ (\nabla A)^t \circ K_V, \nabla A\rangle\Big) = 0. \tag{2.37}$$

From (2.29) and (2.32) we obtain that

$$K_V \circ (\nabla V)^t = \frac{1}{f_V} \begin{pmatrix} f_V^2 - 1 & 0 \\ 0 & 0 \end{pmatrix} \tag{2.38}$$

and thus, the sum of the third and fourth terms of the Hessian is

$$\begin{aligned} f_V \operatorname{tr}((\nabla A)^t \circ \nabla A \circ L_V^{-1}) - \operatorname{tr}((\nabla A)^t \circ K_V \circ (\nabla V)^t \circ \nabla A \circ L_V^{-1}) \\ = \left( f_V - \frac{f_V^2 - 1}{f_V} \right) \alpha + f_V \beta = \frac{\alpha + f_V^2 \beta}{f_V} \end{aligned} \tag{2.39}$$

with

$$\begin{aligned} \alpha &= (\nabla A)_{11}^2\, l_{11} + (\nabla A)_{12}^2\, l_{22} + 2(\nabla A)_{11} (\nabla A)_{12}\, l_{12}, \\ \beta &= (\nabla A)_{21}^2\, l_{11} + (\nabla A)_{22}^2\, l_{22} + 2(\nabla A)_{21} (\nabla A)_{22}\, l_{12}, \end{aligned}$$

where $l_{ij}$ represent the entries of the matrix of $L_V^{-1}$. By direct computation using (2.31) and (2.35)

$$\begin{aligned} f_V^2 \alpha &= E(a)^2 + \frac{1}{r^2} V(a)^2 + \frac{1}{r^2} \Big( E(a) g(\nabla_V V, E) - V(a) g(\nabla_E V, E) \Big)^2 \\ f_V^2 \beta &= \frac{a^2}{r^2} (f_V^2 - 1). \end{aligned} \tag{2.40}$$

If we substitute all the terms in (2.25) by the values obtained in (2.34), (2.37) and (2.39) and we take into account that $\alpha \geq 0$, by the first line of (2.40), then

$$(Hess \mathrm{Vol})_V (A) = \int_M \frac{1}{f_V} \alpha \, dv_g \geq 0$$

with equality if and only if for each unit vector field $E$ orthogonal to $V$ defined in an open set of $M$ the function $a = g(E, A)$ is constant, which is equivalent to $A$ being of constant length. □

## 2.6 Notes

### 2.6.1 Sections That Are Harmonic Maps

The results in Sects. 2.1 and 2.2 appeared in our paper [39] for the tangent and the unit tangent bundles. For the general case of vector bundles and unit vector bundles can be found in our joint work with González-Dávila and Vanhecke [49]. In both papers we deal with a more general case; namely we consider on the manifold $M$ two different Riemannian metrics $\tilde{g}$ and $g$, we view the section as a map $\sigma : (M, \tilde{g}) \to (P, g^S)$ and compute the symmetric 2-form

$$\alpha_\sigma^{\tilde{g}}(X, Y) = \nabla_X^S(\sigma_* \circ Y) - \sigma_*(\tilde{\nabla}_X Y),$$

which is usually known also as the second fundamental form of the map $\sigma$, even when $\tilde{g}$ is different from the induced metric $\sigma^* g^S$; its trace is known as the tension of the map.

The expressions of the second fundamental form and the tension, both for $\sigma \in \Gamma^\infty(P)$ and for $\sigma \in \Gamma^\infty(S^r P)$, are formally equal to those appearing in Proposition 2.1 and in Proposition 2.2. The difference is that now $\tilde{g}$ is the given metric on $M$ and not the metric induced by its immersion $\sigma$ into $(P, g^S)$.

Let's recall that a map between Riemannian manifold is said to be *harmonic* if its tension vanishes. Then $\sigma : (M, \tilde{g}) \to (P, g^S)$ is harmonic if and only if the vector field $X_{(\sigma,\tilde{g})}$ and the section $\eta_{(\sigma,\tilde{g})}$, both defined as in Proposition 2.1, are zero.

Analogously, if $\|\sigma\| = r$ the map $\sigma : (M, \tilde{g}) \to (S^r P, g^S)$ is harmonic if and only if $X_{(\sigma,\tilde{g})} = 0$ and $\eta_{(\sigma,\tilde{g})} \in \mathscr{D}_\sigma$.

It's noteworthy that, although Propositions 2.1 and 2.2 are valid for any $\tilde{g}$, Corollary 2.1 is only true if $\tilde{g} = \sigma^* g^S$ because in general $\alpha_\sigma^{\tilde{g}}(X, Y)$ need not be normal to $\sigma(M)$. For more details the interested reader is referred to [39, 49] and [45].

A particular problem widely studied by different authors since longtime ago consists in considering the conditions for the section $\sigma$ to be a harmonic map from $(M, g)$ to $(P, g^S)$. The first results were due to O. Nouhaud who showed in [76] that on a closed manifold a vector field is a harmonic map into its tangent bundle if and only if it is parallel; Ishihara in [63] obtained this result by giving the expression of the tension. The generalisation to sections of vector bundles over a compact manifold can be found in the paper by Konderak [71].

If we take $\tilde{g} = g$ in Proposition 2.1 we obtain that $\sigma : (M, g) \to (P, g^S)$ is a harmonic map if and only if

$$\sum_{i,j=1}^{n} g(R(E_i, E_j)\sigma, \nabla_{E_j}\sigma)E_i = 0 \qquad \text{and} \qquad \nabla^*\nabla\sigma = 0. \tag{2.41}$$

Therefore if $\sigma$ is a harmonic map then it must be harmonic with respect to the rough Laplacian but not at the inverse. So, a definition similar to Definition 2.2 of minimal sections would be misleading and in many papers (following the papers by Wood [97–99] and [100]) the term *harmonic section* refers, not to sections that are harmonic maps, but to those for which the vertical part of its tension vanishes. Therefore to be a harmonic section in this sense is equivalent to be harmonic with respect to the rough Laplacian. We will see in the next subsection the variational characterisation of this property that will make clear that it's a natural concept.

Nevertheless, this definition does not prevent confusion in the particular case of $\sigma$ being skew-symmetric since then harmonicity is usually referred to $\Delta\sigma = 0$ with $\Delta$ being the Hodge Laplacian, that is related with the rough Laplacian by a Weitzenböck formula.

In particular the more classical definition of a vector field $\sigma$ to be harmonic (see for instance [83] page 167) is that its dual 1-form is Hodge harmonic which is equivalent to

$$\nabla^*\nabla\sigma + \sum_i^n \rho(\sigma, E_i)E_i = 0,$$

where $\rho$ is the Ricci tensor.

It is easy to see that, in contrast with the minimality condition (2.17), both equations in (2.41) are invariant if we change $\sigma$ by a constant multiple $r\sigma$ and consequently in that case it's only necessary to study the harmonicity condition for unit sections which is

$$\sum_{i,j=1}^n g(R(E_i, E_j)\sigma, \nabla_{E_j}\sigma)E_i = 0 \quad \text{and} \quad \nabla^*\nabla\sigma \in \mathscr{D}_\sigma. \tag{2.42}$$

Many authors use the term *unit harmonic section* to refer to unit sections for which only the vertical part of its tension vanishes. In view of the condition above this is equivalent to $\nabla^*\nabla\sigma \in \mathscr{D}_\sigma$ and then with this definition unit harmonic sections need not be harmonic with respect to the rough Laplacian.

### 2.6.2 Sections That Are Critical Points of the Energy Functional

As it occurs with the mean curvature vector field of an immersion, the vanishing of the tension vector field of a map $\varphi : (M, g) \to (N, h)$ has a variational equivalence: A map is harmonic if and only if $\varphi$ is a critical point of the Energy functional for variations with compact support. The Energy being defined as

$$E(\varphi) = \frac{1}{2}\int_M \mathrm{tr}(L_\varphi)\mathrm{dv}_g,$$

where the endomorphism field $L_\varphi$ on $M$ is defined by $(\varphi^*h)(X, Y) = g(L_\varphi(X), Y)$. In the particular case of a section $\sigma : (M, g) \to (P, g^S)$ the Energy has the form

$$E(\sigma) = \frac{1}{2}\int_M (n + \|\nabla\sigma\|^2)\mathrm{dv}_g.$$

This is generally known as the Energy functional on the space of sections of $P$ and, as can be seen in [97], its critical points are characterised by the vanishing of the vertical part of the tension, or equivalently as those that are harmonic with respect to the rough Laplacian. If we consider the restriction of $E$ to unit sections then the Euler-Lagrange equation of the variational problem with this constraint is

$$\nabla^*\nabla\sigma = \|\nabla\sigma\|^2\sigma.$$

So, unit sections that are eigenvalues of the rough Laplacian are critical points of the Energy functional, but a unit section $\sigma$ that is a critical point of the Energy will be an eigenvalue of the rough Laplacian only if in addition $\|\nabla\sigma\|$ is constant.

In the particular case of unit vector fields, G. Wiegmink computed in [94] the first and second variations of a functional which is equal to the Energy, up to constants. The author defined the Total Bending as

$$TB(V) = \frac{1}{(n-1)\mathrm{vol}(S^n)}\int_M \|\nabla V\|^2 \mathrm{dv}_g$$

where $\mathrm{vol}(S^n)$ is the volume of the $n$-dmensional unit sphere. His result for the second variation is the following

**Proposition 2.9** *Let $M$ be a closed manifold. The Hessian of the Energy functional at a critical unit vector field $V$ acting on $A \in \mathscr{D}_V^\perp$ is given by*

$$(Hess E)_V(A) = -\frac{1}{2}\int_M \|A\|^2\|\nabla V\|^2 \mathrm{dv}_g + \frac{1}{2}\int_M \|\nabla A\|^2 \mathrm{dv}_g.$$

It's interesting to compare this simple expression for the second variation of the Energy with the second variation of the Volume (2.25).

In our joint paper with Hurtado [44] we considered the space of timelike unit vector fields on a Lorentzian manifold and defined de Spacelike Energy functional given by the square of the $L_2$ norm of the restriction of the covariant derivative of the vector field to its orthogonal complement. We have computed the first and second variation of this functional which is more adapted to the Lorentzian context.

In this book we are going to deal with the harmonicity properties of tensor fields only when it's convenient for a better understanding of the questions about the minimality. The bibliography about harmonicity of vector fields and sections in general is very abundant and the main reference is the book [34] by S. Dragomir and D. Perrone where the authors also consider different metrics on the bundle and not only the Sasaki metric.

### 2.6.3 Minimal Oriented Distributions

A $q$-dimensional oriented distribution on a manifold $M$ is a section of the Grassmann bundle of the oriented $q$ dimensional subspaces $G_q^o(M) = \cup_{x\in M} G_q^o(T_x M)$. It's not a tensor bundle but a homogeneous bundle and the characterisation of the minimal sections is not a direct consequence of Proposition 2.1.

The characterisation of minimal distributions was obtained in [49] using the following idea of P. M. Chacón, A. M. Naveira and J. M. Weston in [31]: we can identify the bundle $G_q^o(M)$ with the subbundle of the $q$-times contravariant skew-symmetric tensors $\Lambda^q(M)$ consisting in all the decomposable elements of $\Lambda^q(M)$ of norm 1 (see [49] for the details). In particular, one dimensional oriented distributions are then identified with unit vector fields. In [31] this point of view was used to study the harmonicity of distributions.

By abuse of notation, we will write $G_q^o(M) \subset \Lambda^q(M)$ and accordingly we will represent a $q$-dimensional oriented distribution on a manifold $M$ as a unit decomposable section $\sigma$ of $\Lambda^q(M)$. It can be expressed locally as $\sigma = E_1 \wedge \ldots \wedge E_q$, where $\{E_1, \ldots, E_n\}$ is a positive orthonormal local frame such that $\{E_1, \ldots, E_q\}$ span the distribution $\sigma$ and $\{E_{q+1}, \ldots, E_n\}$ span its orthogonal complementary distribution $\sigma^\perp$. We quote the main results, the first of which can be seen as a generalisation of part b) of Theorem 2.4.

**Proposition 2.10** *Let $\sigma$ be a $q$-dimensional oriented distribution on $(M, g)$ and let's represent by $\sigma^\perp$ the $(n-q)$ oriented distribution orthogonal to $\sigma$, the metrics on $M$ induced by both immersions $\sigma : M \to (G_q^o(M), g^S)$ and $\sigma^\perp : M \to (G_{n-q}^o(M), g^S)$ coincide. Moreover, $\sigma$ is minimal if and only if $\sigma^\perp$ is minimal.*

It's easy to see that if $\sigma \in \Lambda^q(M)$ the $q$-times contravariant tensor field $\eta_\sigma$ as defined in Proposition 2.1 is skew-symmetric and then it is a section of $\Lambda^q(M)$.

**Theorem 2.5** *A $q$-dimensional oriented distribution $\sigma$ is minimal if and only if $g(\eta_\sigma, \tilde{\sigma}) = 0$ for all fields of $q$-vectors $\tilde{\sigma}$ that are locally generated by the $q$-vector fields of the form*

$$\sigma_j^a = E_1 \wedge \ldots \wedge E_{a-1} \wedge E_{j+q} \wedge E_{a+1} \wedge E_q$$

*for $a = 1, \ldots, q$, $j = 1, \ldots, n-q$.*

In [54], C. González-Dávila has given an equivalent condition to the minimality in terms of the intrinsic torsion of the associated almost-product structure.

# Chapter 3
# Minimal Vector Fields of Constant Length on the Odd-Dimensional Spheres

In this chapter we will concentrate on the vector fields of constant length on the unit odd-dimensional sphere that are tangent to the fibres of the Hopf fibration. In fact, until now they are the only known examples of smooth minimal vector fields of constant length defined on the whole sphere. It's an interesting open question to determine whether they are the only ones.

Section 3.2 is devoted to study the second variation of the volume functional on these $r$-minimal vector fields. We will show that their stability depends on $r$.

Hopf vector fields can be characterised as the unit Killing vector field of the unit odd-dimensional sphere and this property is useful to generalise the concept of Hopf vector fields to any space form of positive curvature. We will see in Sect. 3.3 that, in contrast with the case of the spheres themselves, in all other quotients the Killing vector fields of constant length are stable critical points of the volume functional.

## 3.1 Minimality of the Hopf Vector Fields

The Hopf fibration, defined by H. Hopf in [60], is the projection from the unit sphere $S^{2m+1} \subset \mathbf{R}^{2m+2} = \mathbf{C}^{m+1}$ onto the complex projective space $\mathbf{C}P^m$. It determines a foliation of $S^{2m+1}$ by great circles and a unit vector field can be chosen as a generator of this distribution. It is given by $H = JN$, where $N$ represents the unit normal to the sphere and $J$ the usual complex structure on $\mathbf{R}^{2m+2}$. So defined $H$ is the standard Hopf vector field.

The following well known result (see for instance [86, pg. 77]) will be used in the sequel

**Proposition 3.1** *If we represent by $g$ the usual metric of the unit sphere and by $\overline{g}$ the Fubini-Study metric of the complex projective space, with sectional curvatures between 1 and 4, then the projection $\pi : (S^{2m+1}, g) \to (\mathbf{C}P^m, \overline{g})$ is a Riemannian submersion with totally geodesic fibres.*



O. Gil-Medrano, *The Volume of Vector Fields on Riemannian Manifolds*,
Lecture Notes in Mathematics 2336, https://doi.org/10.1007/978-3-031-36857-8_3

**Definition 3.1** We will call a *Hopf vector field* any vector field $H$ on $S^{2m+1}$ obtained as $H = JN$ for $J$ a complex structure on $\mathbf{R}^{2m+2}$, that is $J \in End(\mathbf{R}^{2m+2})$ such that $J^t \circ J = \mathrm{Id}$, $J^2 = -\mathrm{Id}$; for $r > 0$ the vector fields of the form $H^r = rJN$ will be called *Hopf vector fields of length r*.

**Proposition 3.2** *The Hopf vector fields of length r on* $S^{2m+1}$ *are r-minimal and their volume is* $\mathrm{Vol}(H^r) = (1+r^2)^m \mathrm{vol}(S^{2m+1})$.

***Proof*** If we denote by $\overline{\nabla}$ the covariant derivative of the Euclidean metric on $\mathbf{R}^{2m+2}$, we have $\overline{\nabla} J = 0$ and then it's easy to compute

$$\begin{aligned}
&\nabla H^r = rJ \quad \text{on} \quad \mathscr{D}_H^{\perp} \quad \text{and} \quad \nabla H^r = 0 \quad \text{on} \quad \mathscr{D}_H, \\
&L_{H^r} = \mathrm{Id} + (\nabla H^r)^t \circ \nabla H^r = (1+r^2)\mathrm{Id} \quad \text{on } \mathscr{D}_H^{\perp} \quad \text{and} \quad L_{H^r} = \mathrm{Id} \quad \text{on} \mathscr{D}_H, \\
&f_{H^r} = (1+r^2)^m, \\
&K_{H^r} = f_{H^r} \nabla H^r \circ L_{H^r}^{-1} = r(1+r^2)^{m-1} J \text{ on } \mathscr{D}_H^{\perp} \text{ and } K_{H^r} = 0 \text{ on } \mathscr{D}_H.
\end{aligned} \tag{3.1}$$

From where the last assertion is an immediate consequence. Let's compute now the divergence of $K_{H^r}$ for which we consider a local orthonormal frame $\{E_i\}_{i=1}^{2m+1}$ with $E_{2m+1} = H$. Since $\nabla_H H = 0$, $K_{H^r}(H) = 0$ and

$$g(\nabla_{E_i} E_i, H) = -g(E_i, \nabla_{E_i} H) = -g(E_i, JE_i) = 0,$$

we have that

$$\begin{aligned}
\nabla^* K_{H^r} &= -\sum_{i=1}^{2m+1} (\nabla_{E_i} K_{H^r}(E_i) - K_{H^r}(\nabla_{E_i} E_i)) \\
&= -r(1+r^2)^{m-1} \sum_{i=1}^{2m} (\nabla_{E_i} JE_i - J(\nabla_{E_i} E_i)).
\end{aligned} \tag{3.2}$$

For $i = 1, \ldots, 2m$,

$$g(\overline{\nabla}_{E_i} JE_i, N) = -g(JE_i, \overline{\nabla}_{E_i} N) = -g(JE_i, E_i) = 0$$

and then

$$\begin{aligned}
\nabla_{E_i} JE_i - J(\nabla_{E_i} E_i) &= \overline{\nabla}_{E_i} JE_i - J(\overline{\nabla}_{E_i} E_i - g(\overline{\nabla}_{E_i} E_i, N)N) \\
&= g(\overline{\nabla}_{E_i} E_i, N)H = -H.
\end{aligned}$$

Using this value in Eq. (3.2)

$$\nabla^* K_{H^r} = 2m(1+r^2)^{m-1} H^r$$

and the result holds from Definition 2.2. □

The Proposition above is a consequence of Proposition 16 of [47] although the proof provided here is different.

**Corollary 3.1** *For a Hopf vector field $H$, the map $H : (S^{2m+1}, g) \to (T^1(S^{2m+1}), g^S)$ is harmonic.*

***Proof*** We have to check that $H$ fulfils the two conditions in (2.42). For any local orthonormal frame $\{E_i\}_{i=1}^{2m+1}$ we have

$$\sum_{i=1}^{2m+1} R(\nabla_{E_i} H, H, E_i) = \sum_{i=1}^{2m+1} (g(\nabla_{E_i} H, E_i)H - g(H, E_i)\nabla_{E_i} H) = -\nabla_H H = 0,$$

which is the first condition in (2.42). That $H$ fulfils the second condition is a direct consequence of Proposition 3.2, since $K_H = 2^{1-m}\nabla H$. □

In [58] D. S. Han and J. W. Yim showed that the only unit vector fields on $S^3$ with the property of being harmonic maps are the Hopf vector fields. The result was improved by H. Gluck and W. Gu in [51] by showing that for a unit vector field on $S^3$ the condition $\sum_{i=1}^{3} R(\nabla_{E_i} V, V, E_i) = 0$ is sufficient to ensure that $V$ is a Hopf vector field.

Recently, I. Fourtzis, M. Markellos and A. Savas-Halilaj have shown in [36] that, for unit vector fields on $S^3$ whose integral curves are great circles, if $V$ is either minimal or a critical point of the Energy then $V$ must be a Hopf vector field. Whether they are the only minimal vector fields on $S^3$ is still an open question.

For higher dimension, D. Perrone showed in [81] that for all odd-dimensional space forms of positive curvature, different from the sphere, the only unit vector fields with the property of being harmonic maps are the projections of the Hopf vector fields.

On the other hand, although we will see in Chap. 5 some examples of minimal vector fields of constant length defined on the sphere minus one or two points, no other examples of minimal smooth vector fields of constant length on the entire sphere have been found, until now, apart from the Hopf vector fields. We are led to the following:

*Conjecture 3.1* The only minimal smooth vector fields of constant length on the odd-dimensional spheres are the Hopf vector fields.

**Definition 3.2** For each unit Hopf vector field $H$ on $S^{2m+1}$ the *Berger metrics* $g_\mu$ associated to $H$ are defined for each $\mu \neq 0$ by

$$g_\mu|_{\mathscr{D}_H^\perp} = g|_{\mathscr{D}_H^\perp}, \qquad g_\mu(H, H) = \mu g(H, H) = \mu$$
$$g_\mu(H, X) = 0 \quad \text{if} \quad X \in \mathscr{D}_H^\perp. \tag{3.3}$$

When $\mu > 0$ the Berger metric is Riemannian and if $\mu < 0$ the metric is Lorentzian and $H$ is timelike. For all $\mu \neq 0$, the map $\pi : (S^{2m+1}, g_\mu) \to (\mathbf{C}P^m, \overline{g})$ is a semi-Riemannian submersion with totally geodesic fibres. The one-parameter family of metrics $(S^{2m+1}, g_\mu)$ is a particular case of a canonical variation of the submersion.

**Corollary 3.2** *The image of the unit Hopf vector fields of* $S^{2m+1}$ *is homothetic to a Berger sphere* $g_\mu$, *with* $\mu = \frac{1}{2}$, *minimally embedded in the Stiefel manifold of orthonormal 2-frames of* $\mathbf{R}^{2m+2}$.

***Proof*** As a direct consequence of the value of $L_{H^r}$ in (3.1), we have that $(H^r)^* g^S = (1+r^2) g_\mu$ with $\mu = \frac{1}{1+r^2}$.

Thus, $H(S^{2m+1})$ is the sphere with the metric $2g_\mu$, for $\mu = \frac{1}{2}$ minimally embedded in $(T^1(S^{2m+1}), g^S)$. It is well known (see for instance [50]) that the unit tangent bundle of $S^{2m+1}$ is isometric to the Stiefel manifold of orthonormal 2-frames of $\mathbf{R}^{2m+2}$ with its usual homogeneous Riemannian structure, obtained by its identification with $SO(2m+2)/SO(2m)$, which ends the proof. □

## 3.2 Study of the Stability of the Hopf Vector Fields

Now we are going to study the stability of the Hopf vector fields by estimating the Hessian of the Volume functional $(Hess\mathrm{Vol})_{H^r}$. We will show that, except for dimension 3, the stability depends of the length.

It is worth mentioning that this section deals with the stability of Hopf vector fields as critical points of the Volume functional in the space of vector fields of constant length $r$, and not with the stability as minimal submanifolds of $T^r(S^{2m+1})$.

**Theorem 3.1** *The Hopf vector fields of length r on* $S^{2m+1}$ *with* $m > 1$ *are unstable for* $r < \sqrt{2m-3}$.

***Proof*** The first step is to show that for each vector field $A \in \mathscr{D}^{\perp}_{H^r} = \mathscr{D}^{\perp}_{H}$

$$\begin{aligned}(Hess\mathrm{Vol})_{H^r}(A) = &-2m(1+r^2)^{m-1}\int_{S^{2m+1}} \|A\|^2 \mathrm{d}v_g \\ &+r^2(1+r^2)^{m-2}\int_{S^{2m+1}} \Big(2m\|JA\|^2 - 2g(JA, \nabla_H A)\Big)\mathrm{d}v_g \\ &+(1+r^2)^{m-2}\int_{S^{2m+1}} \Big(\|\nabla A\|^2 + r^2\|JA\| + r^2\|\nabla_H A\|^2\Big)\mathrm{d}v_g.\end{aligned} \tag{3.4}$$

To do so, we will use the expression of the second variation of the volume functional, formula (2.25), for the particular case of $M = S^{2m+1}$ and $V = H^r$.

Let's compute the first term

$$\begin{aligned}
-\frac{1}{r^2}\|A\|^2\langle K_V, \nabla V\rangle &= -\frac{1}{r^2}\|A\|^2 \mathrm{tr}((\nabla V)^t \circ K_V) \\
&= -\|A\|^2(1+r^2)^{m-1}\mathrm{tr}(J^t \circ J_{|H^\perp}) \\
&= -2m\|A\|^2(1+r^2)^{m-1}. \qquad (3.5)
\end{aligned}$$

For the second term

$$\begin{aligned}
\langle K_V, \nabla A\rangle^2 - \langle K_V \circ (\nabla A)^t \circ K_V, \nabla A\rangle &= (\mathrm{tr}(K_V^t \circ \nabla A))^2 - \mathrm{tr}((K_V^t \circ \nabla A)^2) \\
&= 2\sigma_2(K_V^t \circ \nabla A) \qquad (3.6)
\end{aligned}$$

where $\sigma_2$ represents the second elementary symmetric polynomial function of the endomorphism. On the other hand, if we compute $K_{H^r}^t \circ \nabla A$ in a local orthonormal frame $\{E_1, \ldots, E_{2m}, H\}$ we obtain

$$\begin{aligned}
(K_{H^r}^t \circ \nabla A)(E_i) &= -r(1+r^2)^{m-1}(\nabla_{E_i} JA + g(A.E_i)H) \\
(K_{H^r}^t \circ \nabla A)(H) &= -r(1+r^2)^{m-1}(\nabla_H JA).
\end{aligned}$$

and then

$$\begin{aligned}
\mathrm{tr}(K_{H^r}^t \circ \nabla A) &= -r(1+r^2)^{m-1}\mathrm{tr}(\nabla JA) \\
\mathrm{tr}((K_{H^r}^t \circ \nabla A)^2) &= r^2(1+r^2)^{2(m-1)}\mathrm{tr}((\nabla JA)^2) - 2g(JA, \nabla_H A) \\
2\sigma_2(K_{H^r}^t \circ \nabla A) &= r^2(1+r^2)^{2(m-1)}(2\sigma_2(\nabla JA) + 2g(JA, \nabla_H A)). \qquad (3.7)
\end{aligned}$$

It's well known, see for example [83] pg.170, that for any vector field $X$ on a closed Riemannian manifold $(M, g)$ with Ricci curvature $\rho$ we have the integral equality

$$\int_M \rho(X, X)\mathrm{dv}_g = 2\int_M \sigma_2(\nabla X)\mathrm{dv}_g. \qquad (3.8)$$

Using this equality jointly with (3.6) and (3.7), the expression of the second term of formula (2.25) is

$$r^2(1+r^2)^{m-2}\int_{S^{2m+1}} (2m\|JA\|^2 - 2g(JA, \nabla_H A))\mathrm{dv}_g. \qquad (3.9)$$

The sum of the last two terms of (2.25) can be written as

$$\int_M \langle (f_V \mathrm{Id} - K_V \circ (\nabla V)^t) \circ \nabla A \circ L_V^{-1}, \nabla A\rangle \mathrm{dv}_g$$

but for $V = H^r$, since $\nabla H^r$ is skewsymmetric, $L_{H^r} = \mathrm{Id}-(\nabla H^r)^2$ and it conmutes with $\nabla H^r$. Then

$$f_V \mathrm{Id} - K_V \circ (\nabla V)^t = f_{H^r}(\mathrm{Id} + \nabla H^r \circ L_{H^r}^{-1} \circ \nabla H^r) = f_{H^r} L_{H^r}^{-1} = (1+r^2)^m L_{H^r}^{-1}.$$

Now we only need to compute $\langle L_{H^r}^{-1} \circ \nabla A \circ L_{H^r}^{-1}, \nabla A\rangle$ for which we will use a local orthonormal frame as above and the fact that $g(\nabla_H A, H) = 0$

$$\begin{aligned}
\langle L_{H^r}^{-1} \circ \nabla A \circ L_{H^r}^{-1}, \nabla A\rangle &= \sum_{ij} g((\nabla A \circ L_{H^r}^{-1})(E_i), L_{H^r}^{-1}(E_j)) g(\nabla_{E_i} A, E_j) \\
&\quad + \sum_{i} g((\nabla A \circ L_{H^r}^{-1})(E_i), L_{H^r}^{-1}(H) g(\nabla_{E_i} A, H) \\
&\quad + \sum_{j} g((\nabla A \circ (L_{H^r}^{-1})(H), L_{H^r}^{-1}(E_j)) g(\nabla_H A, E_j) \\
&= (1+r^2)^{-2} \sum_{ij} g(\nabla_{E_i} A, E_j)^2 \\
&\quad + (1+r^2)^{-1} \sum_{i} g(\nabla_{E_i} A, H)^2 \\
&\quad + (1+r^2)^{-1} \sum_{j} g(\nabla_H A, E_j)^2 \\
&= (1+r^2)^{-2} \Big( \|\nabla A\|^2 + r^2 \|JA\| + r^2 \|\nabla_H A\|^2 \Big). \quad (3.10)
\end{aligned}$$

Once computed the second variation of the volume functional at the $r$-minimal vector field $H^r$ it's convenient to write (3.4) in the form

$$\begin{aligned}
&(Hess \mathrm{Vol})_{H^r}(A) \\
&= (1+r^2)^{m-2} \int_{S^{2m+1}} \Big( -2m\|A\|^2 + \|\nabla A\|^2 + r^2 \|JA + \nabla_H A\|^2 \Big) \mathrm{dv}_g \quad (3.11)
\end{aligned}$$

which is obtained by straightforward computation.

For each $a \in \mathbf{R}^{2m+2}$, the vector field $A_a = a - g(a, N)N - g(a, H)H$ is tangent to $S^{2m+1}$ and $A_a \in \mathcal{D}_H^{\perp}$. We compute the Hessian of the Volume functional in the direction of $A_a$. For any vector field $E$ tangent to the sphere and orthogonal to $H$

$$\begin{aligned}
\overline{\nabla}_E A_a &= -g(a, E)N - g(a, N)E - g(a, JE)H - g(a, H)JE \quad \text{and then} \\
\nabla_E A_a &= -g(a, N)E - g(a, JE)H - g(a, H)JE. \quad (3.12)
\end{aligned}$$

On the other hand $\overline{\nabla}_H A_a = -g(a, H)N - g(a, N)H - g(a, JH)H$ and then $\nabla_H A_a = 0$. In a local orthonormal frame $\{E_1, \ldots, E_{2m}, H\}$ such that $E_{2i} =$

$JE_{2i-1}$

$$\|\nabla A_a\|^2 = \sum_{ij} g(\nabla_{E_i} A_a, E_j)^2 + \sum_i g(\nabla_{E_i} A_a, H)^2$$
$$= 2m(g(a, N)^2 + g(a, H)^2) + \sum_i g(A_a, JE_i)^2. \tag{3.13}$$

Since $\|A_a\|^2 = \|JA_a\|^2 = |a|^2 - g(a, N)^2 - g(a, H)^2$, the integrand of (3.11) when we assume that $A = A_a$ is

$$(-2m + 1 + r^2)|a|^2 + (4m - 1 - r^2)(g(a, N)^2 + g(a, H)^2).$$

But

$$\int_{S^{2m+1}} g(a, H)^2 \mathrm{dv}_g = \int_{S^{2m+1}} g(g(a, H)a, H)\mathrm{dv}_g$$
$$= \int_{S^{2m+1}} g(-g(a, H)Ja, N)\mathrm{dv}_g$$
$$= \int_{B^{2m+2}} \mathrm{div}(-g(a, H)Ja)\omega_{2m+2}$$
$$= |Ja|^2 \mathrm{vol}(B^{2m+2}) = \frac{|a|^2}{2m+2}\mathrm{vol}(S^{2m+1})$$

where we have represented by $B^{2m+2} \subset \mathbf{R}^{2m+2}$ the unit ball and by $\omega_{2m+2}$ the Euclidean volume element. Analogously

$$\int_{S^{2m+1}} g(a, N)^2 \mathrm{dv}_g = \frac{|a|^2}{2m+2}\mathrm{vol}(S^{2m+1})$$

and therefore

$$(Hess\mathrm{Vol})_{H^r}(A_a) = (1+r^2)^{m-2}\frac{m}{m+1}|a|^2 \mathrm{vol}(S^{2m+1})(3-2m+r^2). \tag{3.14}$$

Consequently for $m > 1$ and for $r < \sqrt{2m-3}$ the second variation in the direction of $A_a$ with $a \neq 0$ is negative and therefore vector fields $H^r$ are unstable. □

In view of expression (3.11) the sign of the Hessian of the Volume at the Hopf vector fields of the sphere is related with the first eigenvalues of the rough Laplacian acting on vector fields, so we will need the following

**Proposition 3.3** *For the sphere $S^n$ the first eigenvalue of the rough Laplacian acting on vector fields is $\lambda_1^* = 1$ with $\mathscr{E}(\lambda_1^*) = \{\mathrm{grad}(g(a, N))\ ;\ a \in \mathbf{R}^{n+1}\}$ as eigenspace; the second eigenvalue is $\lambda_2^* = n - 1$ with eigenspace $\mathscr{E}(\lambda_2^*)$ being the space of Killing vector fields.*

***Proof*** For the spheres, the spectrum of the Hodge Laplacian acting on $p$-forms is well known (see [37, 64] for example) and in particular for 1-forms the first eigenvalue is $\lambda_1 = n$ and the corresponding eigenspace is given by the differential of the eigenfunctions corresponding to the first non zero eigenvalue of the Laplacian acting on functions $\mathscr{E}(\lambda_1) = \{d(g(a, N))\ ;\ a \in \mathbf{R}^{n+1}\}$. The second eigenvalue is $\lambda_2 = 2(n-1)$ and the eigenspace $\mathscr{E}(\lambda_2)$ is the space of Killing 1-forms.

On the other hand, as can be seen for instance in [83, page 161], for a 1-form $\alpha$ on the sphere (or more generally on a manifold of Ricci curvature $n-1$) the Hodge Laplacian $\Delta$ and the rough Laplacian $\nabla^*\nabla$ are related by

$$\Delta\alpha = \nabla^*\nabla\alpha + (n-1)\alpha.$$

Taking into account the usual duality between 1-forms and vector fields given by the metric $g$ we obtain the result. □

**Proposition 3.4** *The square of the $L_2$-norm of the covariant derivative $\nabla A$ of a vector field $A$ on $S^{2m+1}$ admits the following estimates.*

*(a) For every $A$*

$$\int_{S^{2m+1}} \|\nabla A\|^2 \mathrm{dv}_g \geq \int_{S^{2m+1}} \|A\|^2 \mathrm{dv}_g$$

*and in particular if $A$ is orthogonal to $H$ then*

$$(Hess\mathrm{Vol})_{H^r}(A) \geq \int_{S^{2m+1}} \Big((1-2m)\|A\|^2 + r^2\|JA + \nabla_H A\|^2\Big)\mathrm{dv}_g.$$

*(b) If $A = \sum A^i e_i$ where $\{e_1, \ldots e_{2m+2}\}$ is the standard global orthonormal frame of $\mathbf{R}^{2m+2}$ and $\int_{S^{2m+1}} A^i \mathrm{dv}_g = 0$ for $i = 1, \ldots 2m+2$ then*

$$\int_{S^{2m+1}} \|\nabla A\|^2 \mathrm{dv}_g \geq 2m \int_{S^{2m+1}} \|A\|^2 \mathrm{dv}_g$$

*and if moreover $A$ is orthogonal to $H$ then $(Hess\mathrm{Vol})_{H^r}(A) \geq 0$ for all $r > 0$.*

*(c) If $A$ is orthogonal to $H$ then*

$$\int_{S^{2m+1}} \|\nabla A\|^2 \mathrm{dv}_g \geq \int_{S^{2m+1}} \Big(\|\nabla_H A\|^2 + 3\|A\|^2 + 2mg(\nabla_H A, JA)\Big)\mathrm{dv}_g. \tag{3.15}$$

***Proof*** By Proposition 3.3 and the Rayleigh formula for the rough Laplacian

$$\int_{S^{2m+1}} \|\nabla A\|^2 \mathrm{dv}_g \geq \int_{S^{2m+1}} \|A\|^2 \mathrm{dv}_g, \tag{3.16}$$

for all $A$ and if $A$ is $L_2$-orthogonal to $\mathscr{E}(\lambda_1^*)$ then

$$\int_{S^{2m+1}} \|\nabla A\|^2 dv_g \geq 2m \int_{S^{2m+1}} \|A\|^2 dv_g.$$

For a vector field $A$ orthogonal to $H$ the expression (3.11) of $(Hess\mathrm{Vol})_{H^r}(A)$ combined with (3.16) proves part (a) of the statement and to prove part (b) we only need to take into account that, since $\mathrm{grad}(g(a, N)) = a - g(a, N)N$, a vector field on the sphere is $L_2$-orthogonal to $\mathscr{E}(\lambda_1^*)$ if and only if all its components have vanishing integral.

The proof of part (c) is more involved. To compute $\|\nabla A\|^2$ we take into account that, since $A$ is orthogonal to $H$, then $g(\nabla_H A, H) = 0$.Therefore, if $\{E_i\}_{i=1}^{2m}$ is a local frame of $\mathscr{D}_H^\perp$

$$\|\nabla A\|^2 = \|\nabla_H A\|^2 + \|JA\|^2 + \sum_{i,j=1}^{2m} g(\nabla_{E_i} A, E_j)^2 \tag{3.17}$$

which can be written as $\|\nabla A\|^2 = \|\nabla_H A\|^2 + \|JA\|^2 + \|\tilde{\nabla} A\|^2$, if we represent by $\tilde{\nabla} A$ the restriction of $\nabla A$ to $\mathscr{D}_H^\perp$, both in the domain and in the range. It's easy to check that

$$\|\tilde{\nabla} A\|^2 = \frac{1}{2}\|\tilde{\nabla} A \circ J - J \circ \tilde{\nabla} A\|^2 + \sum_{i=1}^{2m} g(\tilde{\nabla}_{JE_i} A, J\tilde{\nabla}_{E_i} A)$$

$$\|\tilde{\nabla} A\|^2 = \frac{1}{2}\|\tilde{\nabla} A \circ J + J \circ \tilde{\nabla} A\|^2 - \sum_{i=1}^{2m} g(\tilde{\nabla}_{JE_i} A, J\tilde{\nabla}_{E_i} A)$$

and then for $\|\tilde{\nabla} A\|^2$ we have the lower bound

$$\|\tilde{\nabla} A\|^2 \geq |\sum_{i=1}^{2m} g(\tilde{\nabla}_{JE_i} A, J\tilde{\nabla}_{E_i} A)| \tag{3.18}$$

On the other hand, if we assume that $\{E_i\}_{i=1}^{2m}$ is a local $J$-frame of $\mathscr{D}_H^\perp$, i.e. if $E_{i+m} = JE_i$ we are going to prove that

$$2\|A\|^2 = \sum_{i=1}^{2m} g(\tilde{\nabla}_{JE_i} A, J\tilde{\nabla}_{E_i} A) - 2mg(\nabla_H A, JA) + \mathrm{div}(JZ_A) \tag{3.19}$$

where $Z_A = (\nabla A)^t(JA) - g((\nabla A)^t(JA), H)H$. To do so we use the Riemann curvature tensor $R$

$$
\begin{aligned}
\|A\|^2 &= \sum_{i=1}^{m} R(E_i, JE_i, A, JA) \\
&= \sum_{i=1}^{m} \Big( g(\nabla_{JE_i}\nabla_{E_i}A, JA) - g(\nabla_{E_i}\nabla_{JE_i}A, JA) + g(\nabla_{[E_i, JE_i]}A, JA) \Big) \\
&= \sum_{i=1}^{m} \Big( (-g(\nabla_{E_i}A, \nabla_{JE_i}JA) + g(\nabla_{JE_i}A, \nabla_{E_i}JA) \Big) \\
&\quad + \sum_{i=1}^{m} \Big( JE_i(g(\nabla_{E_i}A, JA)) - E_i(g(\nabla_{JE_i}A, JA)) + (\nabla_{[E_i, JE_i]}A, JA) \Big) \\
&= \sum_{i=1}^{2m} g(\nabla_{JE_i}A, \nabla_{E_i}JA) + \mathrm{div}\Big( \sum_{i=1}^{2m} g(\nabla_{E_i}A, JA)JE_i \Big) + g(\nabla_Y A, JA)
\end{aligned}
\tag{3.20}
$$

where for the last equality we have used the well known formula $\mathrm{div}(fX) = f\mathrm{div}(X) + X(f)$ and the notation

$$
Y = \sum_{i=1}^{m} ([E_i, JE_i] - \mathrm{div}(JE_i)E_i + \mathrm{div}(E_i)JE_i).
$$

The first term of (3.20 ) is

$$
\begin{aligned}
\sum_{i=1}^{2m} g(\nabla_{JE_i}A, \nabla_{E_i}JA) &= \sum_{i=1}^{2m} g(\tilde{\nabla}_{JE_i}A, \tilde{\nabla}_{E_i}JA) + \sum_{i=1}^{2m} g(\nabla_{JE_i}A, H)g(\nabla_{E_i}JA, H) \\
&= \sum_{i=1}^{2m} g(\tilde{\nabla}_{JE_i}A, \tilde{\nabla}_{E_i}JA) - \|A\|^2.
\end{aligned}
\tag{3.21}
$$

For the second term, we take into account that the local expression of $Z_A$ in the frame is

$$
Z_A = \sum_{i=1}^{2m} g((\nabla A)^t(JA), E_i)E_i = \sum_{i=1}^{2m} g(JA, \nabla_{E_i}A)E_i. \tag{3.22}
$$

The components of the third term are

$$g(\sum_{i=1}^{m}[E_i, JE_i], H) = \sum_{i=1}^{m}\Big(g(\nabla_{E_i} JE_i, H) - g(\nabla_{JE_i} E_i, H)\Big)$$
$$= \sum_{i=1}^{m}\Big( - g(JE_i, JE_j) - g(E_i, E_i)\Big) = -2m,$$

$$g(\sum_{i=1}^{m}[E_i, JE_i], E_j) = \sum_{i=1}^{m}\Big(g(\nabla_{E_i} JE_i, E_j) - g(\nabla_{JE_i} E_i, E_j)\Big) \tag{3.23}$$
$$= \sum_{i=1}^{m}\Big(g(E_i, \nabla_{E_i} JE_j) + g(JE_i, \nabla_{JE_i} JE_j)\Big) = \mathrm{div}(JE_j),$$

for $j = 1, \ldots, 2m$ and therefore $Y = -2mH$. Now, (3.19) follows from (3.20), (3.21) (3.22) and (3.23).

Part (c) of the statement is obtained after integration by using the estimate (3.18) and equalities (3.17) and (3.19). □

Now we are going to prove that the instability conditions of Theorem 3.1 ($m > 1$ and $r^2 < 2m - 3$) are sharp. We will use the term stable to mean that the Hessian of the volume is positive semidefinite.

**Theorem 3.2** *The Hopf vector fields of length $r$ on $S^{2m+1}$ are stable for $r^2 \geq 2m - 3$. In particular, the Hopf vector fields of length $r$ on $S^3$ are stable for all $r > 0$.*

***Proof*** For simplicity we are going to assume that $H$ is the standard Hopf vector field $H = iN$ and then it is tangent to the fibers of the Hopf fibration $\pi : S^{2m+1} \longrightarrow \mathbf{C}P^m$ that we will represent by $\mathscr{F}(p) = \{e^{it}p \; ; \; t \in [0, 2\pi]\}$. For a general complex structure $J$ on $\mathbf{R}^{2m+2}$ multiplication of a vector $v$ by a complex number $z = x+iy \in \mathbf{C}$ would be given by $zv = xv + yJv$ and in particular $JN$ would be tangent to the fibers $\mathscr{F}(p) = \{\cos tp + \sin tJp \; ; \; t \in [0, 2\pi]\}$.

For a vector field $A$ orthogonal to $H$ we consider the Fourier series of its restriction to the fibers. More precisely, for all $k \in \mathbf{Z}$ we define on the sphere the $\mathbf{R}^{2m+2}$ valued smooth map

$$A_k(p) = \frac{1}{2\pi}\int_0^{2\pi} e^{-ikt} A(e^{it}p)dt.$$

The vectors $e^{-ikt}A(e^{it}p)$ belong to the subspace generated by $\{A(e^{it}p), JA(e^{it}p)\}$ that verifies $\langle\{A(e^{it}p), JA(e^{it}p)\}\rangle \subset \langle\{N(e^{it}p), H(e^{it}p)\}\rangle^{\perp} = \langle\{N(p), H(p)\}\rangle^{\perp}$ and then $A_k$ is a vector field of the sphere orthogonal to $H$.

Since $A$ is smooth the Fourier series converges and for all $t \in [0, 2\pi]$

$$A(e^{it}p) = \sum_{k\in\mathbf{Z}} e^{ikt} A_k(p). \tag{3.24}$$

In particular for $t = 0$ we have $A(p) = \sum A_k(p)$.

By Parseval's identity

$$\int_0^{2\pi} \|A(e^{it}p)\|^2 dt = \sum_{k\in\mathbf{Z}} \int_0^{2\pi} \|A_k(e^{it}p)\|^2 dt. \tag{3.25}$$

By Proposition 3.1, the Hopf fibration is a Riemannian submersion and then for a function $f$ on the sphere

$$\int_{S^{2m+1}} f \mathrm{dv}_g = \int_{\mathbf{C}P^m} \Big( \int_{\mathcal{F}(p)} f dt \Big) \mathrm{dv}_{\overline{g}} \tag{3.26}$$

and then

$$\int_{S^{2m+1}} \|A\|^2 \mathrm{dv}_g = \sum_{k\in\mathbf{Z}} \int_{S^{2m+1}} \|A_k\|^2 \mathrm{dv}_g. \tag{3.27}$$

By construction $(JA)_k = (iA)_k = iA_k = JA_k$. On the other hand $g(A, H) = 0$ implies that $\nabla_H A = \overline{\nabla}_H A$ and then

$$\begin{aligned}(\nabla_H A)(p) &= \frac{d}{dt}_{|t=0} (A(e^{it}p)) = \sum_{k\in\mathbf{Z}} \frac{d}{dt}_{|t=0} \Big(e^{ikt} A_k(p)\Big) = \sum_{k\in\mathbf{Z}} k(JA_k)(p) \\ &= \sum_{k\in\mathbf{Z}} (\nabla_H A_k)(p). \end{aligned} \tag{3.28}$$

For the last equality we only need to take into account that $A_k(e^{it}p) = e^{ikt} A_k(p)$. Using again (3.26)

$$\begin{aligned}\int_{S^{2m+1}} \|\nabla_H A + JA\|^2 \mathrm{dv}_g &= \sum_{k\in\mathbf{Z}} \int_{S^{2m+1}} \|\nabla_H A_k + JA_k\|^2 \mathrm{dv}_g && (3.29) \\ &= \sum_{k\in\mathbf{Z}} \int_{S^{2m+1}} (k+1)^2 \|A_k\|^2 \mathrm{dv}_g. && (3.30)\end{aligned}$$

The map $\overline{\nabla} A$ defined on $TS^{2m+1}$ with values in $\mathbf{R}^{2m+2}$ is related with the endomorphisms field of the sphere $\nabla A$ by $\overline{\nabla}_{X_p} A = \nabla_{X_p} A - g(A, X_p)N_p$ and then $\|\overline{\nabla} A\|^2 = \|\nabla A\|^2 + \|A\|^2$. The next step is to show that

$$\int_{S^{2m+1}} \|\overline{\nabla} A\|^2 \mathrm{dv}_g = \sum_{k\in\mathbf{Z}} \int_{S^{2m+1}} \|\overline{\nabla} A_k\|^2 \mathrm{dv}_g \tag{3.31}$$

which will imply that

$$\int_{S^{2m+1}} \|\nabla A\|^2 \mathrm{dv}_g = \sum_{k\in\mathbf{Z}} \int_{S^{2m+1}} \|\nabla A_k\|^2 \mathrm{dv}_g. \tag{3.32}$$

To do so, for each $p \in S^{2m+1}$ we take an orthonormal basis $\{E_1(p), \dots, E_{2m+1}(p) = H(p)\}$ of $T_p(S^{2m+1})$ and extend it to a reference along the fiber $\mathscr{F}(p)$ by $E_j(e^{it} p) = e^{it} E_j(p)$ then

$$\begin{aligned}
\int_0^{2\pi} \|(\overline{\nabla} A)(e^{it} p)\|^2 dt &= \int_0^{2\pi} \Big( \sum_{j=1}^{2m+1} \|(\overline{\nabla}_{E_j} A)(e^{it} p)\|^2 \Big) dt \\
&= \sum_{j=1}^{2m+1} \sum_{k\in\mathbf{Z}} \int_0^{2\pi} \|(\overline{\nabla}_{E_j} A)_k(e^{it} p)\|^2 dt \\
&= \sum_{j=1}^{2m+1} \sum_{k\in\mathbf{Z}} \int_0^{2\pi} \|(\overline{\nabla}_{E_j} A_k)(e^{it} p)\|^2 dt,
\end{aligned} \tag{3.33}$$

where for the last equality we have used that $(\overline{\nabla}_{E_j} A)_k = \overline{\nabla}_{E_j} A_k$. Indeed, for any vector field along the fiber such that $E(e^{it} p) = e^{it} E(p)$ if $\gamma_{E(p)}(s)$ is a curve with $\gamma_{E(p)}(0) = p$ and $\frac{d}{ds}_{|s=0} \gamma_{E(p)}(s) = E(p)$ then the curve $c(s) = e^{it} \gamma_{E(p)}(s)$ verifies $c(0) = e^{it} p$ and $\frac{d}{ds}_{|s=0} c(s) = E(e^{it} p)$. Therefore

$$\begin{aligned}
(\overline{\nabla}_E A_k)(e^{it} p) &= \frac{d}{ds}_{|s=0} A_k(e^{it} \gamma_{E(p)}(s)) = e^{ikt} \frac{d}{ds}_{|s=0} A_k(\gamma_{E(p)}(s)) \\
&= e^{ikt} \frac{1}{2\pi} \frac{d}{ds}_{|s=0} \int_0^{2\pi} e^{-iku} A(e^{iu} \gamma_{E(p)}(s)) du \\
&= e^{ikt} \frac{1}{2\pi} \int_0^{2\pi} e^{-iku} \frac{d}{ds}_{|s=0} (A(e^{iu} \gamma_{E(p)}(s)) du \\
&= e^{ikt} \frac{1}{2\pi} \int_0^{2\pi} e^{-iku} (\overline{\nabla}_E A)(e^{iu} p) du \\
&= e^{ikt} (\overline{\nabla}_E A)_k(p) = (\overline{\nabla}_E A)_k(e^{ikt} p).
\end{aligned}$$

If we collect Eqs. (3.27), (3.29), and (3.32) we obtain

$$(Hess\text{Vol})_{H^r}(A) = \sum_{k\in\mathbf{Z}} (Hess\text{Vol})_{H^r}(A_k) \tag{3.34}$$

which is the first step of the proof of the Theorem. The second step is to show that $(Hess\text{Vol})_{H^r}(A_k) \geq 0$ for all $k \in \mathbf{Z}$; for $k \neq 0$ it's a direct consequence of part (b) of Proposition 3.4 because in that case

$$\int_0^{2\pi} A_k(e^{it}p)dt = \int_0^{2\pi} e^{ikt} A_k(p)dt = 0.$$

The proof for $A_0$ will follow from part c) of Proposition 3.4 which provides (for vector fields orthogonal to $H$) an estimate of the $L_2$-norm of $\nabla A$ different from the one obtained from the first eigenvalue of the rough Laplacian and which gives a better bound if $\nabla_H A = 0$. Namely, in that case we have

$$\int_{S^{2m+1}} \|\nabla A\|^2 \mathrm{dv}_g \geq 3 \int_{S^{2m+1}} \|A\|^2 \mathrm{dv}_g.$$

Then for the Hessian at $A_0$

$$(Hess\text{Vol})_{H^r}(A_0) \geq (-2m+3+r^2) \int_{S^{2m+1}} \|A\|^2 \mathrm{dv}_g$$

that is greater or equal to 0 under the hypothesis on $r$ and in particular for any length if $m = 1$. □

As a consequence of Theorems 3.1 and 3.2 we can state the following result.

**Corollary 3.3** *For each $m > 1$ there is a particular value of the length $r_0(m) = \sqrt{2m-3}$ such that the Hopf vector fields $H^r$ of length $r$ on $S^{2m+1}$ are stable if and only if $r \geq r_0$.*

Theorems 3.1 and 3.2 were proved in our joint work with Llinares-Fuster [46] and with Borrelli [11], respectively, where unit vector fields on spheres of different radius have been studied instead of vector fields of different length on the unit sphere. The two problems are equivalent (see Lemma 2.2).

By Proposition 3.2 all the Hopf vector fields of length $r$ have the same volume, thus if $A$ is the variational vector field of a variation of $H^r$ through Hopf vector fields of length $r$, then the Hessian acting on $A$ vanishes. Such a variation will be of the form $V(s) = rJ(s)N$ with $J(s)$ being a curve of complex structures such that $J(0) = J$ and then, as it is easy to see, $A$ must be of the form $A = LN$ for $L \in End(\mathbf{R}^{2m+2})$ with $A^t = -A$, $J \circ L = -L \circ J$; therefore the dimension of the space of variational vector fields tangent to variations through Hopf vector fields is $(m+1)^2 + \frac{m(m+1)}{2}$. Consequently, the nullity of the Hopf vector fields is always greater than or equal to this number, which will be referred as the trivial nullity of $H^r$.

In the course of the proof of the Theorem above we have also obtained that

**Proposition 3.5** *For $m > 1$, the non-trivial nullity of the Hopf vector fields of length $r_0(m) = \sqrt{2m-3}$ on $S^{2m+1}$ is $2m+2$.*

*Remark* In [65] D. L. Johnson stated that "the Hopf fibration on the round $S^5$ is not a local minimum of the volume functional" by showing that for any infinitesimal deformation of the unit Hopf vector field the first derivative at $t = 0$ vanishes and that there are deformations for which the second derivative is strictly negative. At a first sight one could think that this result is in contradiction with Theorem 3.2 which for the 5-dimensional sphere shows that Hopf vector fields of length $r$ are stable for $r \geq 1$.

In fact there is a subtlety that was difficult to take into account by the time of the publishing of [65] since our paper [11], where we have showed that the stability of unit vector Hopf fields depend on the radius, has appeared many years later.

In [65] the proof is obtained by using that the Hopf fibration $\pi : S^5 \longrightarrow \mathbf{C}P^2$ is a Riemannian submersion and $\mathbf{C}P^2$ is considered with the usual Fubini-Study metric with holomorphic sectional curvature 1 (or, equivalently with sectional curvatures between $\frac{1}{4}$ and 1). Then the domain of the Hopf fibration has to be $S^5(2)$ the sphere of radius 2 and not the unit sphere (see Proposition 3.1). Therefore the statement of [65] has to be read "The unit Hopf vector field on the round $S^5(2)$ is not a local minimum of the volume functional" that is equivalent to say that the Hopf vector fields of length $r = \frac{1}{\sqrt{2}}$ on $S^5$ are unstable which is compatible with Theorem 3.1 that for the 5-dimensional sphere shows that instability occurs for $r < 1$.

## 3.3 Stability of the Hopf Vector Fields of Odd-Dimensional Space Forms of Positive Curvature

In this section we consider the volume of vector fields on the more general case of the quotients of the odd-dimensional spheres. That the projections of Hopf vector fields of length $r > 0$ are $r$-minimal was shown in [47] by using the characterisation of Hopf vector fields of $S^{2m+1}$ as the unit Killing vector fields. Nevertheless the study of the second variation reveals that with the exception of the spheres themselves the projected vector fields –that we will call also Hopf vector fields by extension– are stable for all $r$. This result was shown by V. Borrelli and H. Zoubir in [13]; we provide a different proof inspired in the proof of first part of Theorem 3.1.

**Proposition 3.6** *A unit vector field $V$ on the sphere $S^{2m+1}(c)$ of curvature $c$ is a Killing vector field if and only if $V = JN$ for $J$ a complex structure on $\mathbf{R}^{2m+2}$.*

***Proof*** Let's recall that a vector field $V$ is Killing if it verifies one of the following equivalente conditions:

- If $\{\varphi_t\}_{t\in\mathbf{R}}$ is the flow of $V$, all the $\varphi_t$ are isometries.
- The Lie derivative verifies $\mathscr{L}_V g = 0$.
- The covariant derivative $\nabla V$ is skewsymmetric.

If $V = JN$ then $\nabla V = \sqrt{c}\, J$ on $\mathscr{D}_V^{\perp}$ and $\nabla V = 0$ on $\mathscr{D}_V$ and then $\nabla V$ is skewsymmetric. Conversely, if $V$ is a unit Killing vector field we can define the map $J : \mathbf{R}^{2m+2} \to \mathbf{R}^{2m+2}$ by

$$J(p) = \|p\|\, V\Big(\frac{1}{\|p\|\sqrt{c}}\, p\Big)$$

for $p \neq 0$ and $J(0) = 0$, which clearly verifies $JN = V$. It's easy to check that if $V$ is Killing the map $J$ is linear, because the flow of $V$ is given by isometries and then $\varphi_t$ is the restriction to the sphere of some $\tilde{\varphi}_t \in O(2m+2)$; the linear map $J$ is in fact an isometry, because $V$ is unit, and $J$ is skewsymmetric since $V$ is tangent to the sphere and then $g(J(p), p) = 0$. Consequently, $J$ is a complex structure of $\mathbf{R}^{2m+2}$ as we wanted to show. □

Unit Killing vector fields on $S^{2m+1}(c)$ will be called also Hopf vector fields.

Any complete manifold $M$ of constant positive curvature $c$ is isometric to the quotient of the sphere $S^n(c)$ by a finite subgroup $G \subset O(n+1)$, of isometries without fixed points, the vector fields of $M$ are the projections of the invariant vector fields of the sphere

$$\Gamma_G^{\infty}(TS^n(c)) = \{X \in \Gamma^{\infty}(TS^n(c))\,;\ \gamma \circ X = X \circ \gamma\ \ \forall \gamma \in G\}.$$

In view of Proposition 3.6, when $n$ is odd, we can define Hopf vector fields on these manifolds as follows

**Definition 3.3** A vector field $V$ on a complete $(2m+1)$-manifold $M$ of constant curvature $c > 0$ will be called a Hopf vector field of length $r$ if one of the two equivalent conditions is verified:

- $V$ is a Killing vector field of constant length $r$.
- $V$ is the projection of a Hopf vector field of length $r$ on $S^{2m+1}(c)$

If $M$ is a projective space, since $G = \{\pm \mathrm{Id}\}$, all the Hopf vector fields of the sphere project to $M$. For general $G$ we know that at least the set of Hopf vector fields is not empty, as deduced from the following classical result of J. A. Wolf in [96].

**Theorem 3.3** *Let $M$ be a complete Riemannian manifold of dimension $2m-1$ and constant curvature $c > 0$. Then $M$ inherits a contact structure from the differential form on $S^{2m+1}(c)$ defined by the restriction of*

$$\omega = \sum_{i=1}^{m}(x^i dx^{m+i} - x^{m+i} dx^i).$$

The short proof is mainly based in the information contained in the last column of the chart in Theorem 7.2.18 of Wolf's book [95], from where the author deduces that every group $G$, as above, is conjugate in $O(2m)$ to a subgroup of $U(m+1)$.

We will recall the concept of contact structure in Sect. 3.4.3, but to proof the Corollary below we only need to take into account that the restriction to the sphere of the 1-form $\omega$ is precisely the 1-form associated, by the metric, with the Hopf field.

**Corollary 3.4** *Every complete odd dimensional manifold of positive constant curvature inherits a Hopf vector field.*

The first eigenvalue of the rough Laplacian acting on vector fields is different for the quotients of the sphere.

**Proposition 3.7** *For $M = S^{2m+1}/G$, $G \neq \{\mathrm{Id}\}$, the first eigenvalue of the rough Laplacian acting on vector fields is $\lambda_1^* = 2m$ with eigenspace $\mathcal{E}(\lambda_1^*)$ being the space of Killing vector fields.*

***Proof*** For such an $M$, the spectrum of the Hodge Laplacian acting on 1-forms is included in the spectrum of the sphere and consists on those $\lambda_i$ such that the space of $G$-invariant eigenforms is different from $\{0\}$; the corresponding eigenspace consists on the projections of the elements of

$$\mathcal{E}^G(\lambda_i) = \{\omega \in \mathcal{E}(\lambda_i)\,;\ \gamma^*\omega = \omega\ \ \forall\gamma \in G\}.$$

As described in Proposition 3.3, $\omega \in \mathcal{E}^G(\lambda_1)$ if an only if $\omega = df_a$ with $f_a = g(a, N)$ and $a \in \mathbf{R}^{2m+2}$, $a \neq 0$. Such a 1-form would be $G$-invariant if and only if $\forall\gamma \in G$ the equality $d(f_a \circ \gamma) = \gamma^* df_a = df_a$ holds, which will imply

$$df_a = \frac{1}{\#(G)} d\Big(\sum_{\gamma\in G} f_a \circ \gamma\Big).$$

But, for every $p \in S^{2m+1}$

$$\sum_{\gamma\in G} f_a(\gamma(p)) = g(a, \sum_{\gamma\in G}\gamma(p)) = 0,$$

since $\sum_{\gamma\in G}\gamma(p)$ is a fixed point for the action of $G$ which is fixed point free, thus $\mathcal{E}^G(\lambda_1)=\{0\}$.

On the contrary, $\mathcal{E}^G(\lambda_2)\neq\{0\}$ because the restriction of the 1-form $\omega$ of Theorem 3.3 is a Killing 1-form, since it is the 1-form associated with the Hopf vector field, and thus it is an element of $\mathcal{E}^G(\lambda_2)$; so, $4m$ belongs to the spectrum and in this case it is the lowest eigenvalue. Now the result is obtained by the same arguments used in Proposition 3.3. □

*Remark* When $M=\mathbf{R}P^n$, it's easy to see that every Killing vector field on $S^n$ is $\mathbf{Z}_2$-invariant thus, the eigenvalue $2(n-1)$ has the same multiplicity in the spectra of both the sphere and the projective space.

Proposition 3.2 can be easily generalised as follows

**Proposition 3.8** *Every Killing vector field $V$ of length $r>0$, on a Riemannian manifold $(M,g)$ of dimension $n=2m+1$ and constant curvature $c$, is $r$-minimal with volume*

$$\mathrm{Vol}(V)=(1+cr^2)^m\mathrm{vol}(M).$$

***Proof*** A vector field as in the statement is a Killing vector field of length $r\sqrt{c}$ on the manifold $(M,cg)$, which has constant curvature equal 1. Therefore we can apply Proposition 3.2 to obtain that $V$ is $r\sqrt{c}$-minimal on $(M,cg)$ and that its volume is

$$\mathrm{Vol}^{\sqrt{c}}(V)=(1+cr^2)^m\mathrm{vol}(M,cg).$$

By Lemma 2.2, $V$ is then $r$-minimal as a vector field on $(M,g)$ and

$$\mathrm{Vol}^{\sqrt{c}}(V)=c^{\frac{n}{2}}\mathrm{Vol}(V),$$

from where the result follows, because $\mathrm{vol}(M,cg)=c^{\frac{n}{2}}\mathrm{vol}(M)$. □

**Theorem 3.4** *Let $M$ be any space form of positive curvature different from the sphere. The Hopf vector fields of length $r$ are stable for all $r>0$.*

***Proof*** We can assume without loss of generality that $M$ is of constant curvature 1, by the arguments in the above Proposition. It's evident that the first part of the proof of Theorem 3.1, and in particular equality (3.11), is valid for $M$ as in the hypothesis. Thus, if $\overline{H}^r$ is a Hopf vector field of length $r$ and $A$ is any vector field on $M$ orthogonal to $\overline{H}^r$ then

$$(Hess\mathrm{Vol})_{\overline{H}^r}(A)\geq(1+r^2)^{m-2}\int_M\Big(-2m\|A\|^2+\|\nabla A\|^2\Big)\mathrm{dv}_g.$$

Consequently, the Hessian is positive semidefinite by Proposition 3.7 and Rayleigh formula. □

## 3.4 Notes

### *3.4.1 Spheres and Their Quotients with Berger Metrics*

In the joint work with Hurtado [45] we have extended the study of the volume of Hopf vector fields to the spheres with a Berger metric. Although in this book we have been dealing only with positive definite metrics, the equations for a vector field to be minimal and to determine a harmonic map can be easily extended to a timelike vector field $V$ on a Lorentzian manifold. The equations only differ in the sign of the terms involving $V$ that in the case of Hopf vector fields don't appear since $H_\mu$ has geodesic flow. In [45] we also study other functionals as the generalised energy and the timelike energy but here we are only to report the results concerning the Riemannian case ($\mu > 0$) and the volume functional. The reader interested in these functionals or in the Lorentzian Berger spheres is referred to [44, 45] and the work [62] by A. Hurtado.

In this subsection, for the Berger sphere $(S^{2m+1}, g_\mu)$, with $\mu > 0$, we will consider the Hopf vector field $H$ used to define $g_\mu$, as in Definition 3.2, so the unit Hopf vector field of $(S^{2m+1}, g_\mu)$ is $H_\mu = \frac{1}{\sqrt{\mu}} JN$.

Using Koszul formula, one obtains the relation of $\nabla$ with $\nabla^\mu$,the Levi-Civita connection of the metric $g_\mu$.

$$\nabla^\mu_H X = \nabla_H X + (\mu - 1)\nabla_X H, \qquad \nabla^\mu_X H = \mu \nabla_X H, \qquad \nabla^\mu_X Y = \nabla_X Y, \tag{3.35}$$

for all $X$, $Y$ in $\mathscr{D}^\perp_H$.

In particular, $H_\mu$ is a unit Killing vector field with totally geodesic flow. We can use the characterisation of the minimality described in Sect. 3.4.2 since the curvature of the Berger spheres is well known. More precisely, the sectional curvature $K_\mu$ of $(S^{2m+1}, g_\mu)$ takes the value

$$K_\mu(\sigma) = 1 + (1 - \mu) g(X, JY)^2, \tag{3.36}$$

if $\sigma \subset H^\perp$ and $\{X, Y\}$ is an orthonormal basis and it takes the value $K_\mu(\sigma) = \mu$, if the plane $\sigma$ contains de vector $H_\mu$.

Since in [45] we have also studied the generalised energy functionals, the minimality has been obtained there from the more general result.

**Proposition 3.9** *For all $\mu, \lambda > 0$, the map $H_\mu : (S^{2m+1}, g_\lambda) \to (T^1(S^{2m+1}), g^S_\mu)$ is harmonic.*

The metric induced by the embedding $H_\mu : S^{2m+1} \to (T^1(S^{2m+1}), g^S_\mu)$ is

$$(H_\mu)^* g^S_\mu = (1 + \mu) g_\lambda$$

where $\lambda = \mu/(1 + \mu)$ and therefore the proposition above implies the minimality.

**Corollary 3.5** *For all $\mu > 0$, the unit Hopf vector field $H_\mu$ is minimal.*

Let's explore now the second variation of the volume at these critical points.

**Proposition 3.10** *If $(2m-2)\mu^3 > \mu + 1$ then the unit Hopf vector field $H_\mu$ of the Berger sphere $(S^{2m+1}, g_\mu)$ is unstable with index at least $2m+2$.*

***Proof*** We only sketch the proof, for the complete details the reader is referred to [45]. Let $H_\mu$ be the Hopf unit vector field on $(S^{2m+1}, g_\mu)$, for each vector field $A$ orthogonal to $H_\mu$ we can show by similar arguments to those used in the proof of Theorem 3.1 that

$$(Hess\mathrm{Vol})_{H_\mu}(A) = (1+\mu)^{m-2}\int_{S^{2m+1}} f(m,\mu,A)\mathrm{dv}_\mu$$

where

$$f(m,\mu,A) = \mu(-2m\mu + 2(1-\mu))\|A\|^2 + \|\nabla^\mu A\|^2 + \mu\|\nabla^\mu_{H_\mu} A + \sqrt{\mu} JA\|^2.$$

As in Theorem 3.1, for each $a \in \mathbf{R}^{2n+2}$, $a \neq 0$, we consider the vector field $A_a \in \mathscr{D}_H^\perp$ given by $A_a = a - g(a,N)N - g(a,H)H$ and we compute the Hessian in its direction obtaining

$$(Hess\mathrm{Vol})_{H_\mu}(A_a) = (1+\mu)^{m-2}\frac{m}{m+1}|a|^2\mathrm{vol}(S^{2m+1}, g_\mu)$$
$$\left((1-2m)\mu(1+\mu) + 2m\mu + 2 + (1+\mu)\frac{(\mu-1)^2}{\mu}\right).$$

The hypothesis on $\mu$ is then equivalent to $(Hess\mathrm{Vol})_{H_\mu}(A_a) < 0$, which ends the proof. □

The following result has been obtained in [45], using the techniques developed in [11] to prove the stability of the Hopf vector fields on round spheres when the radius is big enough.

**Proposition 3.11** *If $(2m-2)\mu^3 \leq \mu + 1$ and $\mu \leq 1$, the unit Hopf vector field $H_\mu$ of the Berger sphere $(S^{2m+1}, g_\mu)$ is stable.*

Propositions 3.10 and 3.11 completely solve the stability question for $m > 1$ since $(2m-2)\mu^3 \leq \mu + 1$ automatically implies $\mu \leq 1$.

**Corollary 3.6** *The unit Hopf vector field $H_\mu$ of the Berger sphere $(S^{2m+1}, g_\mu)$ with $m > 1$ is stable if and only if $(2m-2)\mu^3 \leq \mu + 1$.*

However, for the 3-dimensional case $(S^3, g_\mu)$ the Proposition 3.11 implies that for $\mu \leq 1$ the unit Hopf vector field $H_\mu$ is stable but the hypothesis of Proposition 3.10 never holds. Nevertheless, we have shown in [45] the following result.

**Proposition 3.12** *The unit Hopf vector field $H_\mu$ of the Berger sphere $(S^3, g_\mu)$ with $\mu > 1$ is unstable.*

In this case, the variations responsible for the instability are not those in the direction of the vector fields $A_a$ but those in the direction of the Hopf vector fields orthogonal to $H$. More precisely: after identifying $\mathbf{R}^4$ with the quaternions $\mathbf{H}$, the three unit quaternions $\{i, j, k\}$ give rise to the three Hopf vector fields $H$, $E_1 = jN$ and $E_2 = kN$. If $A$ is an element of the vector space generated by $\{E_1, E_2\}$ then the Hessian of the Volume acting on $A$ is negative.

Propositions 3.11 and 3.12 completely solve the stability question for $m = 1$.

**Corollary 3.7** *The unit Hopf vector field $H_\mu$ of the Berger sphere $(S^3, g_\mu)$ is stable if and only if $\mu \leq 1$.*

As a conclusion, for each dimension $2m + 1$ there is a value $\mu_0(m) \leq 1$ such that $H_\mu$ is stable if and only if $\mu \leq \mu_0(m)$. It's straightforward to obtain the explicit expression of $\mu_0(m)$; in particular, it is strictly less than 1 with the exception of $\mu_0(1) = \mu_0(2) = 1$.

It is worth mentioning that with the Berger metrics with $\mu < 1$ the sphere is isometrically immersed as a geodesic sphere in the complex projective space and that, for $\mu > 1$ it is isometrically immersed as a geodesic sphere of the complex hyperbolic space.

**Open Question** To determine wheter this fact is related with the stability of the Hopf vector fields.

A. Hurtado has studied in [62] the stability of the vector field $rH_\mu$ on the Berger spheres after checking that they are $r$-minimal for all $r > 0$. In particular she has shown that, for the 3-dimensional case, the stability is independent of the length. For $m > 1$ the situation becomes quite complicated and the problem is not completely solved. We state the results obtained in [62], adapted to the point of view of this book of considering vector fields of varying length.

**Proposition 3.13** *Let $(S^{2m+1}, g_\mu)$ be the Berger sphere then the $r$-minimal vector field $rH_\mu$ verifies:*

1. *If $0 < \mu \leq \frac{1}{\sqrt{2m-2}}$, $rH_\mu$ is stable for all $r$.*
2. *If $\frac{1}{\sqrt{2m-2}} < \mu \leq 1$, $rH_\mu$ is stable if and only if $r \geq r_0(m, \mu)$.*
3. *If $\mu \geq \frac{1}{2}\left(1 + \sqrt{\frac{m+1}{m-1}}\right)$, $rH_\mu$ is unstable for all $r$.*

*Here for $\frac{1}{\sqrt{2m-2}} < \mu \leq 1$ we define*

$$r_0(m, \mu) = \sqrt{\frac{(2m-2)\mu^2 - 1}{(2m-2)\mu^2(1-\mu) + \mu}}.$$

**Open Problem** To complete the classification of $rH_\mu$ attending its stability as an $r$-minimal vector field of the Berger $(S^{2m+1}, g_\mu)$ for $m > 1$.

In [13] the authors have studied not only the space forms but also the Riemannian manifolds obtained as a quotient of $(S^{2m+1}, g_\mu)$ by a finite subgroup $G \subset U(m+1)$ of isometries of the Berger sphere, acting freely. We will represent the quotient metric and the unit Hopf vector field by $g_\mu^G$ and $H_\mu^G$, respectively. The stability problem is still open in general, although the following results have been obtained in [13].

**Proposition 3.14** *Let $(S^{2m+1}/G, g_\mu^G)$ be a quotient of the Berger sphere, as described above, different from the sphere. Then*

1. *For all $\mu \leq 1$ the unit Hopf vector field $H_\mu^G$ is stable.*
2. *If $G = \mathbf{Z}_2$ and $m = 2k$, for all $\mu > 1$ the unit Hopf vector field $H_\mu^G$ is unstable.*
3. *If $m > 1$, for $\mu$ large enough the Hopf vector field $H_\mu^G$ is unstable.*

**Open Problem** It would be interesting to find the values of $\mu^G$ for which $H_\mu^G$ is stable; this is an open question even in the case of the projective spaces with $m$ odd. It would be also interesting to give a joint proof of Propositions 3.10 and 3.11 by using the lowest eigenvalues of the rough Laplacian acting on vector fields of the Berger spheres, as we have done in Theorem 3.1 for the round sphere. We recall that this approach has allowed us to give a very simple proof of Theorem 3.4 and could be useful to obtain the results on the quotients; in [62] A. Hurtado also used the lowest eigenvalues of the Laplacian of the Berger metrics acting on functions to study the stability in the Lorentzian case.

### *3.4.2 The Minimality Condition for Unit Killing Vector Fields*

In [47] we have shown that the minimality condition, which is given in terms of the second order differential operator $\nabla^* K_V$, in the particular case of a unit Killing vector field can be written in terms of its first covariant derivative and of the curvature tensor of the manifold and so, the Euler-Lagrange equation reduces to a first order one. Apart from this reduction in the order of the equation, this expression is particularly interesting if we assume certain properties on the curvature of the manifold; for instance, from this expression one can obtain Proposition 3.8 as a corollary.

Let $V$ be a Killing vector field and let's represent by $\rho_V$ the vector field defined in terms of the Ricci tensor by $g(\rho_V, X) = \rho(V, X)$. It's well known (see [83, p. 169]) that $\nabla^*\nabla V = \rho_V$ and therefore, a unit Killing vector field $V$ is a critical point of the Energy if and only if $\rho_V \in \mathcal{D}_V$. This fact was observed in [94].

In the case of the Volumen functional, the differential operator $\nabla^* K_V$ is more complicated but nevertheless, for Killing vector fields, it can be described in terms of the curvature tensor and of the first derivative $\nabla V$.

**Theorem 3.5** *Let $V$ be a unit Killing vector field on a Riemannian manifold $M$ of dimension $n$, then $\nabla^* K_V = f_V \tilde{\rho}_V$, where $\tilde{\rho}_V$ is the vector field defined, in terms of any local orthonormal frame $\{E_j\}_{j=1}^n$, by*

$$g(\tilde{\rho}_V, X) = \sum_{j=1}^{n} R(V, E_j, (L_V^{-1} \circ \nabla V)(E_j), (L_V^{-1} \circ \nabla V)(X))$$

$$+ \sum_{j=1}^{n} R(V, E_j, L_V^{-1}(E_j), L_V^{-1}(X)), \tag{3.37}$$

*for all vector field $X$. Consequently, a unit Killing vector field $V$ is minimal if and only if $\tilde{\rho}_V \in \mathscr{D}_V$.*

***Proof*** Let $V$ be a vector field as in the statement, we know that the rank of $\nabla V$ must be even and for each point of an open dense subset of $M$ we can find a local adapted orthonormal frame

$$\{E_i, E_{i^*}, E_{2k+1}, \ldots, E_n = V\}_{i=1}^k,$$

with $2k$ the rank of $\nabla V$ and such that $\nabla V(E_i) = -\lambda_i E_{i^*}$, $\nabla V(E_{i^*}) = \lambda_i E_i$ for $i \in \{1, \ldots, k\}$ and $\nabla V(E_\alpha) = 0$ for $\alpha \in \{2k+1, \ldots, n\}$.

In such a frame $L_V(E_i) = (1+\lambda_i^2)E_i$, $L_V(E_{i^*}) = (1+\lambda_i^2)E_{i^*}$ for $i \in \{1, \ldots, k\}$ and $L_V(E_\alpha) = E_\alpha$ for $\alpha \in \{2k+1, \ldots, n\}$.

Since, by definition, $K_V = f_V \nabla V \circ L_V^{-1}$ it follows that

$$K_V(E_i) = -f_V \frac{\lambda_i}{1+\lambda_i^2} E_{i^*}, \quad K_V(E_{i^*}) = f_V \frac{\lambda_i}{1+\lambda_i^2} E_i \quad \text{for} \quad i \in \{1, \ldots, k\}$$

and $K_V(E_\alpha) = 0$, for $\alpha \in \{2k+1, \ldots, n\}$.

It's not difficult to see (Lemma 11 of [47]) that for a unit Killing vector field $V$ we have $R(X, V, Y, V) = g(\nabla_X V, \nabla_Y V)$ and then $\lambda_i^2$ is the common value of the sectional curvatures of the planes generated by $V$ and $E_i$ and by $V$ and $E_{i^*}$.

The result can be obtained by showing that for all vector fields $X$

$$g(\tilde{\rho}_V, X) = \frac{1}{f_V} g(\nabla^* K_V, X). \tag{3.38}$$

Let us assume first that $X = E_\alpha$, for $\alpha \in \{2k+1, \ldots, n\}$, then

$$g(\tilde{\rho}_V, E_\alpha) = \sum_{j=1}^{k} \frac{1}{1+\lambda_j^2} \left( R(V, E_j, E_j, E_\alpha) + R(V, E_{j^*}, E_{j^*}, E_\alpha) \right)$$

$$+ \sum_{\beta=2k+1}^{n} R(V, E_\beta, E_\beta, E_\alpha).$$

Here we have used that $L_V^{-1} \circ \nabla V = \nabla V \circ L_V^{-1}$. By the definition of the curvature tensor and the expression of $\nabla V$, it's clear that $R(V, E_\beta, E_\beta, E_\alpha) = 0$ and a direct computation gives $R(V, E_j, E_j, E_\alpha) + R(V, E_{j^*}, E_{j^*}, E_\alpha) = \lambda_j g([E_j, E_{j^*}], E_\alpha)$. Therefore,

$$g(\tilde{\rho}_V, E_\alpha) = \sum_{j=1}^{k} \frac{\lambda_j}{1+\lambda_j^2} g([E_j, E_{j^*}], E_\alpha). \tag{3.39}$$

On the other hand

$$\begin{aligned} g(\nabla^* K_V, E_\alpha) &= -\sum_{i=1}^{n} g(\nabla_{E_i} K_V(E_i), E_\alpha) + \sum_{i=1}^{n} g(K_V(\nabla_{E_i} E_i), E_\alpha) \\ &= f_V \sum_{j=1}^{k} \frac{\lambda_j}{1+\lambda_j^2} g(\nabla_{E_j} E_{j^*} - \nabla_{E_{j^*}} E_j, E_\alpha). \end{aligned} \tag{3.40}$$

Equalities (3.39) and (3.40) prove equality (3.38) for $X = E_\alpha$. The proof for $X = E_i$ and $X = E_{i^*}$, although a little longer, is obtained with similar arguments. The reader can confront Theorem 14 in [47] for details, where a typo is to be corrected in the statement: with the notation used there the statement should read $\omega_V = -f\tilde{\rho}_V$ and not $\omega_V = f\tilde{\rho}_V$. □

The condition in Theorem 3.5 becomes very simple when the dimension of the manifold is 3 or 4. More generally, if for a unit Killing vector field $V$ the rank of $\nabla V$ is 2 then $V$ is minimal if and only if

$$\begin{aligned} &\rho(V, E_1) = \rho(V, E_2) = 0 \qquad \text{and} \\ &\rho(V, E_\alpha) = \frac{-\lambda}{1+\lambda^2} \left( R(E_\alpha, E_1, E_1, V) + R(E_\alpha, E_2, E_2, V) \right). \end{aligned}$$

For a manifold of dimension 3 or 4, as can be seen in [47], the minimality condition becomes a condition of the Ricci tensor and of $V$ and it's not a differential equation.

**Corollary 3.8** *Let $V$ be a unit Killing vector field on a manifold of dimension* 3 *or* 4. *Then $V$ is minimal if and only if $\rho(V, A) = 0$ for all $A \in \mathcal{D}_V^\perp$.*

For $n = 3$, since the curvature tensor is determined by the Ricci tensor, it is easy to conclude that $V$ is minimal if and only if for every $A, B \in \mathcal{D}_V^\perp$, we have $R(A, B, V, X) = 0$ for all vector field $X$.

In [47] we have used Theorem 3.5 to obtain examples of minimal vector fields on the generalised Heisenberg groups $H(1,r)$ and their compact quotients. $H(1,r)$ is the Lie group of dimension $2r+1$ consisting of all real matrices of the form

$$\begin{pmatrix} & & x^1 & z^1 \\ \mathrm{Id}_r & & \vdots & \vdots \\ & & x^r & z^r \\ 0 \cdots 0 & & 1 & y \\ 0 \cdots 0 & & 0 & 1 \end{pmatrix}$$

where $\mathrm{Id}_r$ represents the $r \times r$ identity matrix. A system of global coordinates is given by $\{x^i, z^i, y\,;\, i = 1, \ldots, r\}$.

$H(1,r)$ can be equipped with an invariant metric $g$ for which we have the following orthonormal frame of invariant vector fields:

$$U_i = \frac{\partial}{\partial x^i}, \quad T_i = \frac{\partial}{\partial z^i} \quad \text{for } 1 \leq i \leq r \quad \text{and} \quad W = \sum_{i=1}^{r} x^i \frac{\partial}{\partial z^i} + \frac{\partial}{\partial y}.$$

This construction generalises the classical 3-dimensional Heisenberg group.

If we consider the subgroup $\Gamma(1,r)$ of $H(1,r)$ consisting in those elements of $H(1,r)$ with entries in $\mathbf{Z}$, we can define also the compact manifold $M(1,r) = \frac{H(1,r)}{\Gamma(1,r)}$. The metric in $M(1,r)$ is chosen such that the projection $p : H(1,r) \to M(1,r)$ is a local isometry.

Since $\{U_1, \ldots, U_r, T_1, \ldots, T_r, W\}$ are invariant by the $\Gamma(1,r)$-action we can consider the corresponding frame $\{\tilde{U}_1, \ldots, \tilde{U}_r, \tilde{T}_1, \ldots, \tilde{T}_r, \tilde{W}\}$ in $M(1,r)$ and the results concerning the first frame are also valid for the second one. In particular $T_i$ and $\tilde{T}_i$ are unit Killing vector fields and we have used Theorem 3.5 to show by straightforward computations the following

**Proposition 3.15** *For each $i \in \{1, \ldots, r\}$ $T_i$ (resp. $\tilde{T}_i$) is a minimal vector field of $H(1,r)$ (resp. $M(1,r)$).*

### 3.4.3 Minimality of the Characteristic Vector Field of a Contact Riemannian Manifold

Hopf vector field is the characteristic vector field of the standard Sasakian structure of the odd-dimensional sphere. Using Theorem 3.5 we have shown in [47] that the minimality is a property shared by all such characteristic vector fields.

**Proposition 3.16** *The characteristic vector field of any Sasakian manifold is minimal.*

The result is also true for $K$-contact manifolds (see [55]). Sasakian manifolds are a particular case of contact metric manifolds that are defined as follows:

**Definition 3.4** Let $(M, g)$ be a Riemannian manifold, $\xi$ a unit vector field and $\phi$ an endomorphisms field. Let's denote by $\eta$ the 1-form determined by $\xi$ as $\eta(X) = g(\xi, X)$, we will say that $(M, g, \xi, \phi)$ is a contact metric manifold if for all vector fields $X, Y$

- $\eta(\phi(X)) = 0$,
- $\phi^2(X) = -X + \eta(X)\xi$,
- $g(\phi(X), \phi(Y)) = g(X, Y) - \eta(X)\eta(Y)$,
- $d\eta(X, Y) = g(\phi(X), Y)$.

Then $\xi$ is called the characteristic vector field.

A standard reference for contact geometry is the book by Blair [4].

A contact metric manifold is said K-contact if $\xi$ is a Killing vector field and it is said Sasakian if $(\nabla_X \phi)(Y) = \eta(Y)X - g(X, Y)\xi$. It's easy to se that any Sasakian manifold is K-contact. The choices $\xi = H$ and $\phi = \nabla H$ on the odd-dimensional unit spheres turn them into Sasakian manifolds.

In contact metric geometry, it's a very natural question to study the relation of the properties of the structure with the condition of the characteristic vector field being minimal or a critical point of the energy functional, therefore an important number of works have followed this path. For instance, extending the results in [56], D. Perrone has shown in [79] that the characteristic vector field of a 3-dimensional contact manifold is minimal if and only if the manifold is either Sasakian or a unimodular Lie group equipped with a left-invariant contact metric structure which is not Sasakian. Although the minimality of the characteristic vector field has been studied by several authors, the energy functional has received more attention among the experts in contact geometry, as can be seen in the book by Dragomir and Perrone [34].

The minimality of the characteristic vector field of a K-contact manifold was also obtained by P. Rukimbira in [85] by a different method, namely by showing that the image of the characteristic vector field of a contact metric manifold is a contact invariant submanifold of the unit tangent bundle, endowed with the natural contact structure defined by the metric, and thus it must be minimal.

Let's recall that the characteristic vector field of the natural contact metric structure on $(T^1M, g^S)$ is the geodesic vector field defined as follows: for $u \in T^1_pM$ let $\gamma_u$ be the geodesic passing through $p$ with tangent vector $u$ and let $c(t)$ be the curve on $T^1M$ given by $c(t) = \gamma_u'(t) \in T^1_{\gamma_u(t)}M$ which is a horizontal lift of $\gamma_u$. We define $\Gamma(u) = c'(0)$.

It is easy to see that $\Gamma$ is a unit vector field of $(T^1M, g^S)$ and one can study when it is a unit minimal one. As the covariant derivative of $g^S$ involves deeply the curvature of $g$, the minimality condition produces complicated expressions unless one assumes that the curvature of $M$ has particular properties. So, one can not expect to obtain results valid for a general manifold without curvature assumptions. The best results we know are due to E. Boeckx and L. Vanhecke in [5] that have shown

**Proposition 3.17** *The geodesic vector field $\Gamma : T^1M \to (T^1(T^1M), (g^S)^S)$ of a two-point homogeneous space is minimal. Moreover, if the dimension of $M$ is either 2 or 3 and $\Gamma$ is minimal then $M$ has constant curvature.*

Although in the unit tangent bundle of a Riemannian manifold the only distinguished unit vector field is the geodesic field, as far as the manifold is endowed with an almost hermitian structure $(M, g, J)$, it is possible to define a one parameter family of special horizontal unit vector fields on $T^1M$ by $\Gamma_\alpha = \cos\alpha\, \Gamma + \sin\alpha\, (\Gamma \circ J)$. Furthermore, a vertical unit vector field on the unit tangent bundle is well defined by $\tilde{\Gamma}(u) = (Ju)^{ver}$. In [5] it has been shown that if $(M, g, J)$ is a complex space form, $\tilde{\Gamma}$ and $\Gamma_\alpha$, for all $\alpha$, are minimal vector fields.

In what concerns the second variation of the volume, in [10], V. Borrelli has studied the stability of the characteristic vector field of any Sasakian manifold when one changes the metric by homotethies and has shown that the phenomenon described in Corollary 3.3 is not particular of the sphere but that it also occurs for all Sasakian metrics.

A. Hurtado in [61] has generalised the results in [10] in two senses: on one hand by considering the more general case of K-contact manifolds $(M, g, \xi, \phi)$ and on the other hand by considering on $M$ a two-parameter family of metrics by re-scaling the metric $g$ in the direction of the characteristic vector field by a constant factor $\mu \neq 0$, in the same way described for the Berger metrics on odd-dimensional spheres in Definition 3.2, and then by changing these metrics $g_\mu$ by homotheties, obtaining the metrics $g_\mu^\lambda = \lambda g_\mu$, for every $\lambda > 0$ and the corresponding unit vector fields $\xi_\mu^\lambda$. It has been shown in [61].

**Proposition 3.18** *All the vector fields $\xi_\mu^\lambda$ are minimal. For each $\mu \neq 0$ the set of values $\lambda$ for which the vector field $\xi_\mu^\lambda$ is stable as a minimal unit vector field on $(M, g_\mu^\lambda)$ is an interval.*

### 3.4.4 Minimal Invariant Vector Fields on Lie Groups and Homogeneous Spaces

An important geometrical difference of $S^3$ with the rest of odd dimensional spheres is that it is isometric to a Lie group, more precisely to $SU(2)$ endowed with its natural bi-invariant metric. Moreover, the Hopf vector fields are the invariant vector fields.

This fact led several authors to study the volume functional of vector fields on Lie groups and in particular they have exhibited many examples of invariant minimal unit vector fields.

Let $(G, g)$ be a Lie group endowed with a left-invariant metric, and let $(\mathfrak{g}, \langle, \rangle)$ be its Lie algebra. The condition for a left-invariant unit vector field $V$ to be minimal can be expressed in terms of the corresponding elements $V_e$ and $(\nabla^* K_V)_e$ of $\mathfrak{g}$ and this fact has been used by K. Tsukada and L. Vanhecke in [91] to show very easily the following existence result that we extend to $r$-minimal vector fields.

**Lemma 3.1** *Any odd-dimensional Lie group with left-invariant metric admits $r$-minimal left-invariant unit vector fields.*

***Proof*** Let's consider the map from $\{v \in \mathfrak{g} \ ; \ |v| = r\}$ to $\mathfrak{g}$ given by

$$v \to (\nabla^* K_V)_e - \frac{1}{r^2}\langle(\nabla^* K_V)_e, v\rangle v$$

where $V$ is the left-invariant vector field in $G$ corresponding to $v$. By construction, this map is a vector field on the even dimensional sphere $\{v \in \mathfrak{g} \ ; \ |v| = r\}$ and there exists at least a point $v$ where it vanishes, i. e. such that $(\nabla^* K_V)_e = \frac{1}{r^2}\langle(\nabla^* K_V)_e, v\rangle v$. To conclude we only need to take into account that for any $a \in G$ the equality $(\nabla^* K_V)_a = (L_a)_*((\nabla^* K_V)_e)$ holds. □

The map $V \to \sqrt{\det(L_V)}$, defined on the manifold of unit vector fields and with values in the space of smooth functions, gives rise to a real valued map $f$ defined on the unit sphere of $\mathfrak{g}$. In [91], the authors have found the following equivalent condition to minimality in terms of $f$.

**Proposition 3.19** *Let $(G, g)$ be a Lie group endowed with a left-invariant metric and let $v \in \mathfrak{g}$ be a unit vector. The corresponding left-invariant vector field $V$ is minimal if and only if for all $x \in v^\perp$*

$$df_{|v}(x) = -\mathrm{tr}(\mathrm{ad}_y),$$

*where $y$ is the element of the Lie algebra corresponding to the value at $e$ of the vector field $K_V^t(X)$, with $X$ being the left-invariant vector field corresponding to $x$.*

Let us recall that a Lie group is said unimodular if $\mathrm{tr}(\mathrm{ad}_x) = 0$, for all $x \in \mathfrak{g}$ and that, for a non-unimodular Lie group, the unimodular kernel is the codimension one ideal of $\mathfrak{g}$ defined by

$$\mathfrak{u} = \{x \in \mathfrak{g} \ ; \ \mathrm{tr}(\mathrm{ad}_x) = 0\}.$$

As a clear consequence of Proposition 3.19, the authors of [91] conclude the following result.

**Proposition 3.20** *A left-invariant unit vector field on a unimodular Lie group is minimal if and only if it is a critical point of $f$. In particular, on any such a group there exists at least one minimal left-invariant unit vector field.*

Examples of unimodular Lie groups are the generalised Heisenberg groups (for description and notations see Sect. 3.4.2). Proposition 3.20 was used, in [91], to obtain a full description of their left-invariant minimal unit vector fields. This description completed Proposition 3.15 and a previous result of C. González-Dávila and L. Vanhecke in [55], which shows that the unit invariant vector field $W$ is minimal. Also in [55], it has been proved that the unit vector fields $U_i$ are minimal.

As another application of Propositions 3.19 and 3.20, and using the Milnor classification (see [75]) of invariant metrics on Lie groups as a key ingredient, the authors give in [91] a complete classification of left-invariant minimal unit vector fields on any 3-dimensional Lie group. Let's mention that the only three cases for which every invariant unit vector field is minimal are $\mathbf{R}^3$ (whose invariant vector fields are parallel), $SU(2)$ with the bi-invariant metrics (that are isometric to 3-dimensional spheres and whose invariant vector fields are the Hopf vector fields) and the Heisenberg group $H(1, 2)$.

It is noteworthy that, although it is not possible to use Proposition 3.20 if the group is not unimodular, the computations needed to apply Proposition 3.19 are not very complicated due to the dimension restriction. For arbitrary dimension, C. González-Dávila and L. Vanhecke have shown in [55] that unit left-invariant vector fields orthogonal to the unimodular kernel are minimal in any dimension.

The same authors have studied in [57] the stability of the invariant minimal unit vector fields on 3-dimensional Lie groups classified in [91]. As pointed out in the paper, all those that have been shown to be stable are Killing vector fields.

**Open Problem** As far as we know, the only results concerning the second variation of the volume of vector fields on Lie groups are those contained in [57]. It would be interesting to have a deeper knowledge of the stability of the different examples of minimal vector fields contained in the papers reported in this subsection.

The study of the volume of unit vector fields on a compact semisimple Lie group has been done by M. Salvai in [87], using the characterisation of the minimality in Proposition 3.19 and the structure and properties of semisimple Lie groups, in a development where the roots of the Lie algebra play a central role. To describe the results we need some definitions; for more details the reader is referred for instance to [59].

Let $\mathfrak{g}$ be the Lie algebra of a compact semisimple Lie group $G$. Let $\mathfrak{t}$ be a maximal abelian subalgebra, $\Delta$ the corresponding root system, $C$ a Weyl chamber and $\Phi$ the associated basis of $\Delta$. This means that $C = \{v \in \mathfrak{t} \ ; \ \alpha(v) > 0, \forall\alpha \in \Phi\}$ and that the closure of the Weyl chamber is the simplex $\overline{C} = \{v \in \mathfrak{t} \ ; \ \alpha(v) \geq 0, \forall\alpha \in \Phi\}$.

Given $\alpha \in \Phi$, there is a unique $v_\alpha \in \mathfrak{t}$ such that $\beta(v_\alpha) = 0$ for all $\beta \neq \alpha$, $\alpha(v_\alpha) > 0$ and $|v_\alpha| = 1$. Each vector $v_\alpha$ is a vertex of the simplex

$$\overline{C}^1 = \overline{C} \cap \{v \in \mathfrak{g} \ ; \ |v| = 1\}.$$

This vector is maximal singular (i. e. its $Ad(G)$-orbit has dimension strictly less than the orbit of any other unit vector in a neighbourhood). Each maximal singular unit vector belongs to the $Ad(G)$-orbit of exactly one of this vertex.

The main result in [87] is the following Theorem.

**Theorem 3.6** *On a compact semisimple Lie group, for any maximal singular unit vector, the corresponding left invariant and right invariant unit vector fields are minimal.*

*Moreover, if $\alpha \neq \beta$ and the unit vector fields corresponding to $v_\alpha$ and $v_\beta$ have the same volume, then there is $v$ in the edge joining $v_\alpha$ with $v_\beta$ (i. e. $\gamma(v) = 0$ for all $\gamma \neq \alpha$, $\gamma \neq \beta$) and different from $v_\alpha$ and $v_\beta$, such that the corresponding left invariant and right invariant unit vector fields are minimal.*

In [87] two vector fields $V_1$, $V_2$ are said to be equivalent if and only if there exists an isometry $\varphi$ such that $V_1 \circ \varphi = \varphi_* \circ V_2$. It's an immediate consequence of the definition that equivalent vector fields have the same volume and that $V_1$ is minimal if and only if $V_2$ is minimal. From the Theorem above, the author obtains a lower bound for the number of nonequivalent minimal vector fields.

In the same paper, the author computes the expression of the volume of the left invariant (and the right invariant) vector field $V$ determined by a unit $v \in \mathfrak{g}$, in terms of the root system, obtaining:

$$\mathrm{Vol}(V) = \mathrm{vol}(G) \prod_{\phi \in \Delta^+} \left(1 + \frac{1}{4}\phi(v)^2\right).$$

Let us end this subsection by the extension of these results on Lie groups to the more general situation of homogeneous spaces. In our joint work with González-Dávila and Vanhecke [48] we have constructed several examples of invariant minimal unit vector fields on homogeneous spaces and in most cases we have determined the complete set of invariant minimal unit vector fields. To do so, we have derived a criterion for the minimality of such a vector field using the framework of homogeneous structures [89] and infinitesimal models [90].

Let $(M = G/G_0, g)$ be a homogeneous Riemannian manifold with reductive decomposition $\mathfrak{g} = \mathfrak{m} \oplus \mathfrak{g}_0$ and let $\tilde{\nabla}$ be the adapted canonical connection. We will consider the associated homogeneous structure $S = \nabla - \tilde{\nabla}$ and its trace $\eta = \sum_i S_{E_i} E_i$ where $\{E_i\}$ is an orthonormal basis of $\mathfrak{m}$.

The associated infinitesimal model is $\mathfrak{M} = (\mathfrak{m}, \tilde{T}, \tilde{R}, <, >)$, where $\tilde{T}$ and $\tilde{R}$ are the torsion and curvature of $\tilde{\nabla}$. The subspace of invariant vectors of $\mathfrak{m}$ is defined as

$$\mathrm{Inv}(\mathfrak{M}) = \bigcap_{x,y \in \mathfrak{m}} \mathrm{Ker}\, \tilde{R}_{x,y}.$$

**Proposition 3.21** *Let $(M = G/G_0, g)$ be a homogeneous Riemannian manifold with reductive decomposition $\mathfrak{g} = \mathfrak{m} \oplus \mathfrak{g}_0$, a left-invariant unit vector field $V$ is minimal if and only if*

$$df_{|v}(x) = - < \eta, (K_V^t(X))_e >,$$

*for all $x \in \mathrm{Inv}(\mathfrak{M}) \cap v^\perp$.*

Among the different classes of homogeneous structures given in [89], we can find the hyperbolic spaces which are in fact isometric to Lie groups. More precisely, if we consider the Poincaré half-space model $H^n = \{y \in \mathbf{R}^n \ ; \ y_1 > 0\}$ with the metric of constant negative curvature $-c^2$ given by

$$g = \frac{1}{(cy_1)^2} \sum_{i=1}^{n} (dy_i)^2,$$

$(H^n, g)$ is isometric to the subgroup of $(Gl(n, \mathbf{R}), \frac{1}{c^2}\langle\ ,\ \rangle)$ consisting in all matrices of the form

$$a = \begin{pmatrix} e^{y_1}\mathrm{Id}_{n-1} & A \\ 0 & 1 \end{pmatrix}$$

where $A = (y_2, \ldots, y_n)$. Then an orthonormal frame of left-invariant vector fields is of the form

$$Y_i = ce^{y_1}\frac{\partial}{\partial y_i}, \quad \text{for } 2 \leq i \leq n \quad \text{and} \quad Y_1 = c\frac{\partial}{\partial y_1}.$$

In [48] we have proved the following result.

**Proposition 3.22** *On the hyperbolic space of curvature $-c^2$, if $n = 2$ then every unit left-invariant vector field is minimal but if $n \geq 3$ the unit vector fields $\pm Y_1$ are the only minimal unit invariant vector fields.*

Let's remark that since the result is valid for all $c$ then it's also valid for vector fields of any constant length.

On naturally reductive homogeneous manifolds, invariant vector fields are Killing and this fact enables us to show that if for one of these manifolds the space $\mathrm{Inv}(\mathfrak{M})$ is 1-dimensional then the unit generators are minimal. The same conclusion is also true for unit invariant $V$ such that $\nabla V$ is of rank 2. With all these properties we determine all the invariant minimal unit vector fields on naturally reductive homogeneous spaces of dimension $\leq 5$.

Another interesting class of homogeneous structures $S$ is defined by the conditions $\eta = 0$ and $< S_X Y, Z > + < S_Z X, Y > + < S_Y Z, X >= 0$. The full classification of connected, complete and simply connected Riemannian manifolds of dimension 3 and 4 which admit a non trivial such structure was obtained by O. Kowalski and F. Tricerri in [74]. In [48], we combine this classification with Proposition 3.21 to determine all the invariant minimal unit vector fields on homogeneous spaces of dimension 3 and 4.

Examples of minimal foliations on homogeneous Riemannian manifolds obtained by using the minimality characterisation by the extrinsic torsion can be seen in [54].

### 3.4.5 Examples Related with Complex and Quaternionic Structures

The Hopf unit vector fields on odd-dimensional spheres are a particular case of a general construction on any orientable real codimension 1 submanifold $(M, g)$ of a Kähler manifold $(\bar{M}, \bar{g}, J)$: we can define on $M$ the unit vector field $\xi = JN$, where $N$ represents the unit normal to the hypersurface. In fact $\xi$ is the characteristic vector field of an almost-contact metric structure on $M$ if we define the endomorphism field on the submanifold by the expression $\phi(X) = JX - \bar{g}(JX, N)N$.

The submanifold $M$ is said a *Hopf hypersurface* if $\xi$ determines a principal direction, that is if $S_\xi = \alpha\xi$, where $S$ represents the shape operator. In order to characterise the minimality of $\xi$, K. Tsukada and L. Vanhecke have defined in [93] for a Hopf hypersurface of dimension $2n+1$ the function

$$\tilde{h} = \sum_{i=1}^{2n} \operatorname{arc\,cot} \lambda_i,$$

where $\{\lambda_i \;\; ; \;\; i = 1, \ldots, 2n\}$ are the principal curvatures of $\xi^\perp$. This function is defined only in the open dense set where the multiplicities of the principal curvatures are locally constant. They have shown the following result

**Proposition 3.23** *The Hopf vector field of a Hopf hypersurface of a complex space form is minimal if and only if* $\tilde{h}$ *is constant.*

Among the examples described by the authors thanks to this characterisation, one can find the particular case of the Hopf hypersurfaces whose principal curvatures are constant; this class contains for instance tubes about complex submanifolds in complex space forms which have been classified in [69] and [2].

A real hypersurface $M$ is said ruled if $\xi^\perp$ defines a foliation with totally geodesic leaves. The authors show in [93] that the Hopf vector field of a minimal real ruled hypersurface of a complex space form of non-zero curvature is minimal. Since examples of this kind of hypersurfaces were known previously, the result provides examples of minimal Hopf vector fields

A surprising result, also in [93], is the following:

**Proposition 3.24** *The Hopf vector field of a Hopf hypersurface of a complex space form of constant holomorphic sectional curvature equal* 4 *is minimal.*

The same authors describe in [92] examples of minimal unit vector fields on hypersurfaces of the Grassmannian of complex two-planes $\mathrm{Gr}_2(\mathbf{C}^{m+2})$. This manifold carries a Kähler structure $J$ and a quaternionic Kähler structure $\mathcal{J}$. For $m \geq 3$, they consider real hypersurfaces with normal bundle $M^\perp$ such that $J(M^\perp)$ and $\mathcal{J}(M^\perp)$ are invariant under the action of the shape operator. They proved that the corresponding unit Hopf vector fields of these hypersurfaces are minimal and analogously when $Gr_2(\mathbf{C}^{m+2})$ is replaced by its non compact dual space $Gr_2(\mathbf{C}^{m+2})^*$.

If we consider the sphere $S^{4m-1} \subset \mathbf{R}^{4m} = \mathbf{H}^m$ and we represent by $\{J_1, J_2, J_3\}$ its usual quaternionic structure, the three Hopf vector fields $H_a = J_a N$, $a = \{1, 2, 3\}$ define a 3-vector field $\sigma = H_1 \wedge H_2 \wedge H_3$ on $S^{4m-1}$, which determines the smooth 3-dimensional distribution $\mathscr{H}$, generated by $\{H_1, H_2, H_3\}$.

This distribution, tangent to the fibers of the Hopf fibration $\pi : S^{4m-1} \to \mathbf{H}P^{m-1}$, is known as the Hopf distribution of the $(4m-1)$-dimensional sphere and in [49] we have shown that it is minimal.

**Proposition 3.25** *The 3-dimensional Hopf distribution of $S^{4m-1}$ is minimal.*

***Proof*** Let $\{E_i\}_{i=1}^n$, $n = 4m-1$, be a positive oriented local frame such that $E_a = H_a$, for $a \in \{1, 2, 3\}$, and $\{E_i\}_{i=4}^n$ span $\mathscr{H}^\perp$. By Theorems 2.1 and 2.5, in order to prove that $\sigma$ is minimal we need to show that $g(\nabla^* K_\sigma, \tilde{\sigma}) = 0$, for every 3-vector field $\tilde{\sigma}$ of the form

$$\sigma_j^1 = E_j \wedge H_2 \wedge H_3, \qquad \sigma_j^2 = H_1 \wedge E_j \wedge H_3, \qquad \sigma_j^3 = H_1 \wedge H_2 \wedge E_j,$$

for $j = 4, \ldots, n$.

To compute $K_\sigma$ we will use the well known relations

$$\begin{aligned} &\nabla_{H_1} H_2 = -\nabla_{H_2} H_1 = -H_3, \quad \nabla_{H_2} H_3 = -\nabla_{H_3} H_2 = -H_1, \\ &\nabla_{H_3} H_1 = -\nabla_{H_1} H_3 = -H_2. \end{aligned}$$

Moreover, as we have seen in the proof of Proposition 3.2, for all $a \in \{1, 2, 3\}$ and $X \in \mathscr{H}^\perp$

$$\begin{aligned} &\nabla_X H_a = J_a \quad \text{and} \quad \nabla_{H_a} H_a = 0, \\ &\nabla_X X \in \mathscr{H}^\perp \quad \text{and} \quad \nabla_X (J_a X) - J_a(\nabla_X X) = -H_a \end{aligned} \tag{3.41}$$

As a consequence, if $X \in \mathscr{H}$ then $\nabla_X \sigma = 0$ and for $X \in \mathscr{H}^\perp$

$$\nabla_X \sigma = J_1 X \wedge H_2 \wedge H_3 + H_1 \wedge J_2 X \wedge H_3 + H_1 \wedge H_2 \wedge J_3 X. \tag{3.42}$$

Then, it is easy to see that $L_\sigma = \mathrm{Id}$ on $\mathscr{H}$ and $L_\sigma = 4\mathrm{Id}$ on $\mathscr{H}^\perp$. Thus, $K_\sigma$ is equal to $\nabla\sigma$, up to a constant factor, and we only need to show that $g(\nabla^* \nabla\sigma, \tilde{\sigma}) = 0$. For a unit $X \in \mathscr{H}^\perp$, let's compute $\nabla_X$ of the first term of $\nabla_X \sigma$ in (3.42), using (3.41).

$$\begin{aligned} \nabla_X (J_1 X \wedge H_2 \wedge H_3) &= \nabla_X (J_1 X) \wedge H_2 \wedge H_3 + J_1 X \wedge J_2 X \wedge H_3 \\ &\quad + J_1 X \wedge II_2 \wedge J_3 X \\ &= J_1(\nabla_X X) \wedge H_2 \wedge H_3 - \sigma + J_1 X \wedge J_2 X \wedge H_3 \\ &\quad + J_1 X \wedge H_2 \wedge J_3 X. \end{aligned}$$

If we take, analogously, the covariant derivative of the other two terms of $\nabla_X\sigma$ and make the addition, it's easy to check that for all $\sigma_j^a$, as above, and unit $X \in \mathscr{H}^\perp$

$$g(\nabla_X\nabla_X\sigma - \nabla_{\nabla_X X}\sigma, \sigma_j^a) = 0.$$

The result is now obtained by taking into account that, in the chosen local frame,

$$\nabla^*\nabla\sigma = -\sum_{j=4}^{n}\left\{\nabla_{E_j}\nabla_{E_j}\sigma - \nabla_{\nabla_{E_j}E_j}\sigma\right\}.$$

□

In [49], we have also extended this result to more general situations coming from 3-Sasakian and quaternionic geometry.

In [77] L. Ornea and L. Vanhecke have studied the minimality of some vector fields and distributions that appear in a natural way in the context of locally conformal Kähler and hyperkähler manifolds.

# Chapter 4
# Vector Fields of Constant Length of Minimum Volume on the Odd-Dimensional Spherical Space Forms

For a Riemannian manifold with finite volume, parallel vector fields are exactly the minimisers of the volume of vector fields, the minimum being equal to the volume of the manifold. For a manifold admitting smooth vector fields of constant length but not parallel ones, like the unit odd-dimensional round sphere, and the other compact manifolds of positive constant curvature, one can consider for $r > 0$ the questions of computing the infimum of the volume of smooth vector fields of length $r$ and of determining if this value is attained by some smooth vector fields.

This problem was proposed by H. Gluck and W. Ziller in [52] and solved in the case of $r = 1$ on the 3-dimensional sphere. The result has been extended to vector fields of any constant length $r > 0$ and to any complete 3-dimensional manifolds of positive constant curvature. Section 4.1 is devoted to the proof.

In our joint paper with A. Hurtado [45] we solved the problem in the case of the 3-dimensional sphere with a Berger metric $g_\mu$ with $\mu < 1$. Although in the book we are mainly concentrating on the results concerning constant curvature spaces, we consider that this case, studied in the second section, is interesting because the proof is based on the properties of the global orthonormal frame of $S^3$ obtained from the quaternionic structure of $\mathbf{R}^4$ and will help to understand how special the 3-dimensional case is.

For general odd-dimensional spheres there is a lower bound of the volume functional that was quoted in [52] for the unit vector fields and extended to compact manifolds of constant curvature by F. G. B. Brito, P. M. Chacón and A. M. Naveira in [21]. The proof of this result is in Sect. 4.3.

The chapter ends with a section devoted to the study of the asymptotical behaviour, when $r$ tends to 0 or to $\infty$, of the real function $r \to \mathrm{Vol}(rV)$ for a fixed unit vector field $V$. In particular we are interested in the information we can get from it about the infimum of the volume of smooth vector fields of constant length.



O. Gil-Medrano, *The Volume of Vector Fields on Riemannian Manifolds*,
Lecture Notes in Mathematics 2336, https://doi.org/10.1007/978-3-031-36857-8_4

## 4.1 Hopf Vector Fields as Volume Minimisers in the 3-Dimensional Case

The important paper [52], where the authors show that on the unit $S^3$ the unit vector fields of minimum volume are Hopf vector fields and no others, has served as inspiration of a wide research work. The authors use the fact that the unit tangent manifold $T^1S^{2m+1}$ with the Sasaki metric is isometric to the Stiefel manifold of orthogonal 2-frames of $\mathbf{R}^{2m+2}$ with the homogeneous metric resulting from the diffeomorphism with $SO(2m+2)/SO(2m)$. They find a calibration that in dimension 3 calibrates exactly the images of Hopf vector fields but for greater dimension the calibrated submanifold does not arise from a vector field and then the method can't be extended to higher dimensional spheres.

The proof we give here is completely different and valid for vector fields of any constant length as well as for manifolds of any positive constant curvature; it's worth mentioning that $T^rS^{2m+1}$ with the Sasaki metric for $r \neq 1$ is not isometric to the standard Stiefel manifold and the proof in [52] is not automatically valid. The result has appeared in our article [43] for unit vector fields; at time of publication, we were not aware that the same result, proved with different methods, had been previously published by D. Perrone in [80]. Here, for the sake of completeness, we extend the proof to vector fields of any constant length, which is a simple exercise in virtue of Lemma 2.2.

The first part of the proof is to show that, if $M$ is a complete manifold of positive constant curvature, the minimum of the volume of vector fields of length $r$ is attained by the Killing vector fields; the argument is quite direct and based in Theorem 3 of the paper [16] by F. Brito. The complicated part is to show that the only minimisers are the Killing vector fields; to obtain the result we will use the fact that minimisers of the volume must be $r$-minimal vector fields and the properties of the first two eigenvalues of the rough Laplacian (see Propositions 3.3 and 3.7).

For an endomorphism $A$ of $\mathbf{R}^n$ we will represent by $\sigma_k(A)$ the symmetric functions of $A$; i. e. the homogenous invariant polynomials of degree $k$ in the entries of $A$ such that the characteristic polynomial is given by

$$\det(A - t\mathrm{Id}) = \sum_{i=0}^{n} (-1)^i \sigma_{n-i}(A) t^i. \tag{4.1}$$

If we consider the matrix of $A$ in any basis, $\sigma_k(A)$ is the sum of all its diagonal minors of order $k$.With this notation $\det(\mathrm{Id} + A) = \sum_{i=0}^{n} \sigma_i(A)$.

**Theorem 4.1** *The volume of a vector field $V$ of length $r > 0$ on a complete 3-manifold $M$ with constant curvature $c > 0$ verifies $\mathrm{Vol}(V) \geq (1 + cr^2)\mathrm{vol}(M)$ with equality if and only if $V = rH$ for $H$ a Hopf vector field.*

***Proof*** For $M = S^3(c)/G$, let's first show that $(1 + cr^2)\mathrm{vol}(M)$ is a lower bound. If $V$ is a vector field with $\|V\| = r$, since $g(\nabla_X V, V) = 0$ for all vector field $X$, we can define $P$ as the restriction of $\nabla V$ to the subbundle $\mathscr{D}_V^\perp$ and we have that

$$\sigma_3((\nabla V)^t \circ \nabla V) = \det((\nabla V)^t \circ \nabla V) = \det(\nabla V)^2 = 0$$

$$\sigma_2((\nabla V)^t \circ \nabla V) = \sigma_2(P^t \circ P) + \sum_{j=1}^{2} \left( \|\nabla_V V\|^2 \|\nabla_{E_j} V\|^2 - g(\nabla_V V, \nabla_{E_j} V)^2 \right)$$

$$\sigma_1((\nabla V)^t \circ \nabla V) = \sigma_1(P^t \circ P) + \|\nabla_V V\|^2$$

where $\{E_1, E_2\}$ is a local orthonormal frame of $\mathscr{D}_V^\perp$. Then

$$\begin{aligned} \det(L_V) &= \sigma_2((\nabla V)^t \circ \nabla V) + \sigma_1((\nabla V)^t \circ \nabla V) + 1 \\ &\geq \sigma_2(P^t \circ P) + \sigma_1(P^t \circ P) + 1 = \det(\mathrm{Id} + P^t \circ P), \end{aligned}$$

where equality holds if and only if $\nabla_V V = 0$.

Let $\mu_1^2, \mu_2^2$ with $\mu_1 \geq 0, \mu_2 \geq 0$ be the eigenvalues of the positive semidefinite symmetric endomorphism $P^t \circ P$. By definition

$$\det(\mathrm{Id} + P^t \circ P) = \mu_1^2 \mu_2^2 + \mu_1^2 + \mu_2^2 + 1$$

and then $\det(\mathrm{Id} + P^t \circ P) \geq (1 + \mu_1 \mu_2)^2$ with equality if and only if $\mu_1 = \mu_2$. Finally

$$f_V = \sqrt{\det(L_V)} \geq (1 + \mu_1 \mu_2) \geq 1 + \sigma_2(P) = 1 + \sigma_2(\nabla V) \tag{4.2}$$

where the second inequality is an equality if and only if $\sigma_2(P) = \det P \geq 0$. Now the lower bound of the volume is a consequence of the integral formula, see for example [83, pg. 170],

$$\int_M \rho(X, X) \mathrm{dv}_g = 2 \int_M \sigma_2(\nabla X) \mathrm{dv}_g \tag{4.3}$$

where $X$ is any vector field and $\rho$ is the Ricci curvature, that under the hypothesis verifies $\rho(X, X) = 2cg(X, X)$. Then by (4.2) and (4.3)

$$\mathrm{Vol}(V) \geq \mathrm{vol}(M) + \int_M \sigma_2(\nabla V) \mathrm{dv}_g = (1 + cr^2) \mathrm{vol}(M).$$

It follows from Proposition 3.8 that for $V = rH$ the equality holds.

To show the unicity of Hopf vector fields as minimisers, it will be enough to show that the result is true for all $r > 0$ when $c = 1$, if we take into account Lemma 2.2.

Let's assume that $V$ with $\|V\| = r$ realises the lower bound that in turn implies, as we have seen above, that $\nabla_V V = 0$, $P^t \circ P = \mu^2 \mathrm{Id}$, $\det P \geq 0$, then

$$\int_M (1+\mu^2) \mathrm{dv}_g = \mathrm{Vol}(V) = (1+r^2)\mathrm{vol}(M)$$

and

$$\int_M \|\nabla V\|^2 \mathrm{dv}_g = \int_M 2\mu^2 \mathrm{dv}_g = 2r^2 \mathrm{vol}(M) = 2\int_M \|V\|^2 \mathrm{dv}_g.$$

If $G \neq \{\mathrm{Id}\}$, by Proposition 3.7, the first eigenvalue of the rough Laplacian is $\lambda_1^* = 2$ and the above equality implies that $V$ is a minimiser of the Rayleigh quotient

$$R(V) = \frac{\int_M \|\nabla V\|^2 \mathrm{dv}_g}{\int_M \|V\|^2 \mathrm{dv}_g},$$

therefore $V$ must be an eigenvector of $\lambda_1^*$, as can be seen for instance in [86, p. 266], and consequently it is a Killing vector field and necessarily it is of the form $V = rH$.

If $M$ is the sphere, by Proposition 3.3, $\lambda_2^* = 2$ and thus $R(V) = \lambda_2^*$. In this case, to conclude that $V$ must be an eigenvector of the second eigenvalue of the Laplacian, we need to show that it is $L_2$-orthogonal to the vector fields $\mathrm{grad} f_a$ with $f_a = g(a, N)$ for $a \in \mathbf{R}^4$, $a \neq 0$, which are the eigenvectors of the first eigenvalue of the rough Laplacian, as described in Proposition 3.3.

Being a minimiser, $V$ must be $r$-minimal and then $\nabla^* K_V = \frac{1}{r^2}\langle K_V, \nabla V\rangle V$ by (2.24). Let's recall that $\nabla V$ also verifies the conditions $\nabla_V V = 0$ and $P^t \circ P = \mu^2 \mathrm{Id}$. It is easy to see that $K_V = \nabla V$ and so the minimality condition can be written as

$$\nabla^* \nabla V = \frac{1}{r^2}\|\nabla V\|^2 V = \frac{2\mu^2}{r^2} V. \tag{4.4}$$

Then we have

$$\begin{aligned} \int_{S^3} g(V, \mathrm{grad} f_a)\mathrm{dv}_g &= \int_{S^3} g(V, \nabla^*\nabla \mathrm{grad} f_a)\mathrm{dv}_g = \int_{S^3} g(\nabla^*\nabla V, \mathrm{grad} f_a)\mathrm{dv}_g \\ &= \frac{2}{r^2}\int_{S^3} \mu^2 g(V, \mathrm{grad} f_a)\mathrm{dv}_g. \end{aligned} \tag{4.5}$$

On the other hand, by Lemma 4.1 below

$$\int_{S^3} (r^2+\mu^2) V(f_a)\mathrm{dv}_g = 0$$

which is in contradiction with (4.5), unless

$$\int_{S^3} g(V, \operatorname{grad} f_a)\mathrm{dv}_g = 0.$$

Consequently, $V$ is an eigenvector of $\lambda_2^*$ and it must be a Killing vector field, as we wanted to show. □

Now we are going to show the technical lemma used in the proof of the theorem; although we only use its version for $m = 1$, we are sure that it is interesting for the reader to have the general result for odd-dimensional spheres which is not in [43].

**Lemma 4.1** *Let $V$ be a vector field of length $r > 0$ on $S^{2m+1}$ whose integral curves are geodesics and such that $P$, the restriction of $\nabla V$ to the subbundle $\mathcal{D}_V^\perp$, verifies $P^t \circ P = \mu^2 \mathrm{Id}$ for a nonnegative function $\mu$. If $V$ is $r$-minimal then* $\operatorname{div}((\mu^2 + r^2)V) + (m-1)V(1+\mu^2) = 0$ *and* $(m-1)Z(1+\mu^2) = 0$ *for all vector field $Z \in \mathcal{D}_V^\perp$. In the particular case of $m = 1$ the vector field $(r^2 + \mu^2)V$ has vanishing divergence.*

***Proof*** Since $V$ is $r$-minimal the vector field $X_V$ defined in Proposition 2.1 must vanish and by the hypotheses on $\nabla V$, if $\{E_i\}_{i=1}^{2m+1}$ is an orthonormal local frame with $E_{2m+1} = \frac{1}{r}V$ then $\{\tilde{E}_i\}_{i=1}^{2m+1}$ where $\tilde{E}_i = \frac{1}{\sqrt{1+\mu^2}}E_i$ and $\tilde{E}_{2m+1} = E_{2m+1}$ is a local frame orthonormal with respect to the metric $\tilde{g}$. By Koszul formula, for $i = 1, \ldots, 2m$

$$\tilde{g}(\tilde{\nabla}_V V, \tilde{E}_i) = \tilde{g}([\tilde{E}_i, V], V) = g([\tilde{E}_i, V], V) = 0$$

and then $\tilde{\nabla}_V V = 0$ and the expression of $X_V$ is

$$\begin{aligned} X_V &= \frac{1}{1+\mu^2}\sum_j^{2m}\Big(\nabla_{E_j}E_j - \tilde{\nabla}_{E_j}E_j - R(V, \nabla_{E_j}V, E_j)\Big) \\ &= \frac{1}{1+\mu^2}\sum_j^{2m}\Big(\nabla_{E_j}E_j - \tilde{\nabla}_{E_j}E_j\Big) + \frac{\operatorname{div} V}{1+\mu^2}V. \end{aligned}$$

If $E = E_j$, $j = 1, \ldots, 2m$, and $Z$ is any vector field, again by Koszul formula

$$\begin{aligned} g(\nabla_E E - \tilde{\nabla}_E E, Z) &= g(\nabla_E E, Z) - \tilde{g}(\tilde{\nabla}_E E, L_V^{-1}(Z)) \\ &= Eg(E, Z) + g([Z, E], E) \\ &\quad - E\tilde{g}(E, L_V^{-1}(Z)) + \frac{1}{2}L_V^{-1}(Z)\tilde{g}(E, E) \\ &\quad - \tilde{g}([L_V^{-1}(Z), E], E) \end{aligned}$$

$$= -g(\nabla_E Z, E) + g(\nabla_E L_V^{-1}(Z), L_V(E)) + \frac{1}{2} L_V^{-1}(Z)(1+\mu^2).$$

Therefore

$$g(X_V, V) = \frac{1}{1+\mu^2} \sum_j^{2m} \Big( -g(\nabla_{E_j} V, E_j) + g(\nabla_{E_j} L_V^{-1}(V), L_V(E_j)) \Big) + \frac{m}{1+\mu^2} V(\mu^2) + \frac{r^2 \mathrm{div} V}{1+\mu^2}$$

$$= \frac{1}{1+\mu^2} \Big( (\mu^2 + r^2)\mathrm{div} V + m V(\mu^2) \Big)$$

$$= \frac{1}{1+\mu^2} \Big( \mathrm{div}((\mu^2 + r^2) V) + (m-1) V(\mu^2) \Big).$$

$$g(X_V, E_i) = \frac{1}{1+\mu^2} \sum_j^{2m} \Big( -g(\nabla_{E_j} E_i, E_j) + g(\nabla_{E_j} L_V^{-1}(E_i), L_V(E_j)) \Big) + \frac{m}{1+\mu^2} L_V^{-1}(E_i)(1+\mu^2)$$

$$= \frac{1}{1+\mu^2} \sum_j^{2m} (1+\mu^2) E_j\Big(\frac{1}{1+\mu^2}\Big) g(E_i, E_j) + \frac{m}{(1+\mu^2)^2} E_i(\mu^2)$$

$$= \frac{m-1}{(1+\mu^2)^2} E_i(\mu^2)$$

and the result holds. □

The similar result to the one in the theorem above also holds for the energy functional.

**Proposition 4.1** *On a complete 3-manifold $M$ with constant curvature $c > 0$, for all unit vector field $V$*

$$\int_M \|\nabla V\|^2 \mathrm{dv}_g \geq 2c\mathrm{vol}(M)$$

*with equality if and only if $V = H$ for $H$ a Hopf vector field.*

***Proof*** With the same notation used in Theorem 4.1, and following also [16] for the proof of the estimate part, we have that for a unit vector field $V$

$$\|\nabla V\|^2 = \sigma_1((\nabla V)^t \circ \nabla V) = \sigma_1(P^t \circ P) + \|\nabla_V V\|^2$$

with equality if and only if $\nabla_V V = 0$. Now we take into account that for a $k \times k$ matrix $P$

$$(k-1)\sigma_1(P^t \circ P) = (k-1)\sum_{ij}^{k} P_{ij}^2 = \sum_{i<j}\left((P_{ii} - P_{jj})^2 + (P_{ij} + P_{ji})^2\right)$$

$$+(k-2)\sum_{i\neq j}^{k} P_{ij}^2 + 2\sigma_2(P). \qquad (4.6)$$

Since $P$ is a $2 \times 2$ matrix $\sigma_1(P^t \circ P) \geq 2\sigma_2(P)$ with equality if and only if $P_{12} = -P_{21}$ and $P_{11} = P_{22}$ which implies that $P^t \circ P = \mu^2\mathrm{Id}$ with $\mu = P_{11}$. The lower bound follows now from (4.3) and the equality when $V = H$ follows from Proposition 3.8. The unicity part is a consequence of the proof of the theorem above. □

## 4.2 Hopf Vector Fields on 3-Dimensional Spheres with the Berger Metrics

We have seen in Sect. 3.4.2 that for the Berger spheres $(S^3, g_\mu)$ the Hopf vector fields of constant length are stable critical points if and only if $\mu \leq 1$. We will end this section by showing that as in the case of the round sphere ($\mu = 1$) also for $\mu < 1$ they are the only minimisers of the volume. On the contrary, for $\mu > 1$, we will give examples of vector fields of constant length with volume less than the volume of the Hopf vector fields; which existence we know in advance due to the instability.

The proof makes use of the same arguments of our proof in [45] of the corresponding result for unit vector fields. The unicity part is very simple and different from the proof of Theorem 4.1 above for the round sphere, due to the values of the Ricci tensor in the case of the Berger spheres.

The example is interesting to highlight the rich structure of $S^3$ which is not shared by the spheres of higher odd dimension. In particular, since $S^3 \subset \mathbf{R}^4$ and if we identify $\mathbf{R}^4$ with the quaternions $\mathbf{H}$, we can define on $S^3$ a global $g$-orthonormal frame $\{H = J_0 N, E_1 = J_1 N, E_2 = J_2 N\}$ where $\{J_0 = J, J_1, J_2\}$ denote the three standard complex structures defining the quaternionic structure of $\mathbf{R}^4$. Then $\{H_\mu, E_1, E_2\}$ is a $g_\mu$-orthonormal frame.

**Proposition 4.2** *Let $(S^3, g_\mu)$ be the three-dimensional Berger sphere. If $\mu < 1$ the volume of a vector field $V$ of length $r$ verifies*

$$\mathrm{Vol}(V) \geq \mathrm{Vol}(rH_\mu) = (1 + r^2\mu)\mathrm{vol}(S^3, g_\mu)$$

*with equality if and only if $V = \pm rH^\mu$.*

*If $\mu > 1$, for all vector fields $X$ in the 2-dimensional space generated by $\{E_1, E_2\}$ with $\|X\| = r$ we have*

$$\mathrm{Vol}(X) = \sqrt{(1 + r^2\mu)^2 + 4r^2(1 + r^2\mu)\frac{1-\mu}{\mu}}\mathrm{vol}(S^3, g_\mu) < \mathrm{Vol}(rH_\mu).$$

***Proof*** Let's first compute the volume of $rH_\mu$. It's well known that

$$\nabla_H H = \nabla_{E_1} E_1 = \nabla_{E_2} E_2 = 0 \qquad \nabla_{E_1} E_2 = -\nabla_{E_2} E_1 = -H$$
$$\nabla_H E_1 = -\nabla_{E_1} H = -E_2 \qquad \nabla_{E_2} H = -\nabla_H E_2 = -E_1,$$

then, using (3.35), the covariant derivative of the Berger metric verifies

$$\begin{aligned} &\nabla^\mu_{H_\mu} H_\mu = \nabla^\mu_{E_1} E_1 = \nabla^\mu_{E_2} E_2 = 0 && \nabla^\mu_{E_1} E_2 = -\nabla^\mu_{E_2} E_1 = -\sqrt{\mu}\, H_\mu \\ &\nabla^\mu_{H_\mu} E_1 = \frac{\mu - 2}{\sqrt{\mu}} E_2 && \nabla^\mu_{E_1} H_\mu = \sqrt{\mu}\, E_2 \\ &\nabla^\mu_{H_\mu} E_2 = -\frac{\mu - 2}{\sqrt{\mu}} E_1 && \nabla^\mu_{E_2} H_\mu = -\sqrt{\mu}\, E_1, \end{aligned} \tag{4.7}$$

from where $\det(L_{rH_\mu}) = (1 + r^2\mu)^2$ and

$$\mathrm{Vol}(rH_\mu) = (1 + r^2\mu)\mathrm{vol}(S^3, g_\mu). \tag{4.8}$$

For a vector field $X = a_1E_1 + a_2E_2$ with $a_i \in \mathbf{R}$ and $a_1^2 + a_2^2 = r^2$ by using the expression of $\nabla^\mu$ in (4.7) we obtain by straightforward computation

$$\det(L_X) = (1 + r^2\mu)(1 + r^2\frac{(\mu - 2)^2}{\mu}) = (1 + r^2\mu)^2 + 4r^2(1 + r^2\mu)\frac{1-\mu}{\mu}$$

and if $\mu > 1$ by comparing with (4.8) it's clear that $\mathrm{Vol}(X) < \mathrm{Vol}(rH_\mu)$.

With the same arguments used in the first steps of the proof of Theorem 4.1, for any vector field $V$

$$\mathrm{vol}(V) \geq \mathrm{vol}(S^3, g_\mu) + \frac{1}{2}\int_{S^3} \rho_\mu(V, V)dv_\mu.$$

From (3.36) it's easy to see that if $g(V, H) = 0$

$$\rho_\mu(H_\mu, H_\mu) = 2\mu, \quad \rho_\mu(V, H_\mu) = 0, \quad \rho_\mu(V, V) = 2(2-\mu)g(V, V) \tag{4.9}$$

and consequently if $\mu < 1$ and $V$ is of length $r$ then

$$\rho_\mu(V, V) \geq \rho_\mu(rH_\mu, rH_\mu) = 2r^2\mu$$

with equality if and only if $V = \pm rH_\mu$ which jointly with (4.8) finish the proof. □

**Open Question** It would be interesting to find the infimum of the volume of vector fields of constant length for the 3-dimensional spheres with $\mu > 1$. A proof similar to the one provided in Proposition 4.2 for $\mu < 1$ gives that if $\mu > 1$ and $V$ is of length $r$ then $\mathrm{vol}(V) \geq (1 + r^2(2-\mu))\mathrm{vol}(S^3, g_\mu)$ but in [45] we have shown that this lower bound is never attained. Besides, the bound is relevant only for $\mu < 2$.

## 4.3 Lower Bound of the Volume of Vector Fields of Constant Length

In [52] the authors give a lower bound of the volume of unit vector fields on $S^{2m+1}$; namely $\mathrm{Vol}(V) \geq c(m)\mathrm{vol}(S^{2m+1})$ with

$$c(m) = \sum_{k=0}^{m} \frac{\binom{m}{k}^2}{\binom{2m}{2k}}. \tag{4.10}$$

A different expression for $c(m)$, useful while comparing the volume of a given vector field with the bound, was pointed out in [78]. The first step is to write

$$c(m) = \binom{2m}{m}^{-1} \sum_{k=0}^{m} \binom{2k}{k}\binom{2(m-k)}{m-k}. \tag{4.11}$$

If we consider the binomial series expansion for any real number $\alpha$ which is valid for $|x| < 1$

$$\frac{1}{(1-x)^\alpha} = \sum_{k=0}^{\infty} \frac{(\alpha+k-1)(\alpha+k-2)\cdots\alpha}{k!}x^k,$$

for the particular case of $\alpha = \frac{1}{2}$ and $x = 4y$ we obtain that for all $y$ with $|y| < \frac{1}{4}$

$$\frac{1}{\sqrt{(1-4y)}} = \sum_{k=0}^{\infty} \frac{(k-1/2)(k-3/2)\cdots 1/2}{k!} 4^k y^k$$
$$= \sum_{k=0}^{\infty} \frac{(2k-1)(2k-3)\cdots 3}{k!} 2^k y^k = \sum_{k=0}^{\infty} \binom{2k}{k} y^k$$

from where

$$\frac{1}{(1-4y)} = \sum_{m=0}^{\infty} \sum_{k=0}^{m} \binom{2k}{k} \binom{2(m-k)}{m-k} y^m. \tag{4.12}$$

But if we use the binomial series expansion formula with $\alpha = 1$ and $x = 4y$ what we get is

$$\frac{1}{(1-4y)} = \sum_{m=0}^{\infty} 4^m y^m \tag{4.13}$$

and then (4.11), (4.12) and (4.13) give that

$$c(m) = \binom{2m}{m}^{-1} 4^m \tag{4.14}$$

The method for showing the volume estimate in [52] has been to use the fact that $T^1S^{2m+1}$ is isometric to $SO(2m+2)/SO(2m)$ with the homogeneous metric and to find a calibrating $(2m+1)$-form whose explicit formula appeared in [53]. This result was generalised in [21] to unit vector fields on any manifold of constant curvature with a different method which is basically the one we will follow here. Previously we need some technical results.

**Lemma 4.2** *Let $A$ be an endomorphism of $\mathbf{R}^n$ and let $\{\mu_1, \mu_2 = \overline{\mu}_1, \ldots, \mu_{2s-1}, \mu_{2s} = \overline{\mu}_{2s-1}, \mu_{2s+1}, \ldots \mu_n\}$ be the roots of its characteristic polynomial, that we assume to be complex numbers from 1 to $2s$. Let's consider $A$ expressed in a base in which it has a normal Jordan form and let's represent by $A^{j_1,\ldots,j_k}_{i_1,\ldots,i_k}$ the $k \times k$ submatrix of the matrix of $A$ consisting in the $j_1, \ldots, j_k$ columns and the $i_1, \ldots, i_k$ rows; we will write $A^{j_1,\ldots,j_k}$ for the corresponding diagonal submatrix. Then*

$$\det(A^{j_1,\ldots,j_k}) \le |\mu_{j_1}| \cdots |\mu_{j_k}| \tag{4.15}$$
$$\sum_{i_1<\ldots<i_k} \det(A^{j_1,\ldots,j_k}_{i_1,\ldots,i_k})^2 \ge |\mu_{j_1}|^2 \cdots |\mu_{j_k}|^2. \tag{4.16}$$

***Proof*** Let's first consider the diagonal minors. For $j_1 < \ldots < j_k$, if the indices in odd position $j_1, j_3, \ldots j_{\bar{k}} \leq 2s - 1$ are odd, the indices in even position are $j_2 = j_1 + 1, j_4 = j_3 + 1, \ldots, j_{\bar{k}+1} = j_{\bar{k}} + 1$ and $j_{\bar{k}+2} > 2s$, we have

$$\det(A^{j_1,\ldots,j_k}) = |\mu_{j_1}| \cdots |\mu_{j_{\bar{k}+1}}| \, \mu_{j_{\bar{k}+2}} \cdots \mu_{j_k}.$$

If the condition above is fulfilled only for a subset of indices $\{j_{\bar{s}} < \ldots < j_k\}$ and we write $\mu_j = a_j + ib_j$ then it's not difficult to check that

$$\det(A^{j_1,\ldots,j_k}) = a_{j_1} \cdots a_{j_{\bar{s}-1}} |\mu_{j_{\bar{s}}}| \cdots |\mu_{j_{\bar{k}+1}}| \, \mu_{j_{\bar{k}+2}} \cdots \mu_{j_k} \tag{4.17}$$

from where we get (4.15).

On the other hand, if $j_1 < \ldots < j_h < j_{\bar{s}}$ the submatrix consisting in these $h$ columns of $A$ is of the form

$$\begin{pmatrix} a & 0 & \cdots & 0 \\ -b & 0 & \cdots & 0 \\ 1 & a & \cdots & 0 \\ 0 & -b & \cdots & 0 \\ 0 & 1 & \cdots & 0 \\ \vdots & & & \vdots \\ 0 & \cdots & \cdots 1 & a \\ 0 & \cdots & \cdots 0 & -b \end{pmatrix}$$

where for simplicity we have assumed that is included in one of the Jordan blocks corresponding to a complex root $\mu = a + ib$.

We have seen in (4.17) that $\det(A^{j_1,\ldots,j_h})^2 = a^{2h}$ and if we take $i_1 = j_1 + 1, i_2 = j_2, \ldots, i_h = j_h$ then we have $\det(A^{j_1,\ldots,j_h}_{i_1,\ldots,i_h})^2 = b^2 a^{2(h-1)}$; in fact this is the common value of the $h$ minors obtained with rows $i_1, \ldots, i_h$ where just one $j$ is changed by $j + 1$. It's also immediate that if $i_1 = j_1 + 1, i_2 = j_2 + 1, i_3 = j_3 \ldots, i_h = j_h$ then $\det(A^{j_1,\ldots,j_h}_{i_1,\ldots,i_h})^2 = b^4 a^{2(h-2)}$ which is the common value of the $\binom{h}{2}$ minors obtained with rows where just two of the $j$'s are changed by their consecutive indices, and so on. Then the sum of all these terms is equal to $|\mu|^{2h}$ that jointly with (4.17) give us (4.16). □

**Lemma 4.3** *Let's denote by $\sigma_k$ the symmetric polynomial of degree $k$ of the diagonal matrix of entries $\{a_1, \ldots, a_{2m}\}$, then*

$$\sum_{k=0}^{2m} \frac{1}{\binom{2m}{k}} \sigma_k^2 \geq \Big( \sum_{k=0}^{m} \frac{\binom{m}{k}}{\binom{2m}{2k}} \sigma_{2k} \Big)^2,$$

*with equality if and only if $a_1 = \cdots = a_{2m}$.*

***Proof*** The polynomials $\sigma_k$ verify the Newton inequality

$$\sigma_k^2 \geq \frac{\binom{2m}{k}^2}{\binom{2m}{k-1}\binom{2m}{k+1}}\sigma_{k+1}\sigma_{k-1}$$

from where we can deduce that if $2h < k < 2m - 2h$ then

$$\sigma_k^2 \geq \frac{\binom{2m}{k}^2}{\binom{2m}{k-2h}\binom{2m}{k+2h}}\sigma_{k+2h}\sigma_{k-2h}. \tag{4.18}$$

It's convenient to write

$$\left(\sum_{k=0}^{m} \frac{\binom{m}{k}}{\binom{2m}{2k}}\sigma_{2k}\right)^2 = \sum_{k=0}^{m}\sum_{i=0}^{k} \frac{\binom{m}{i}\binom{m}{k-i}}{\binom{2m}{2i}\binom{2m}{2k-2i}}\sigma_{2i}\sigma_{2k-2i},$$

where for each $k$ the corresponding term of the sum collects all the terms of degree $2k$ (i. e. those consisting in the product of $2k$ $a_i$'s). The inequality in the statement is obtained by comparing, for each $k$, the term of degree $2k$ by using (4.18). The reader is referred to [21] for more details. □

**Theorem 4.2** *The volume of a vector field $V$ of length $r > 0$ on a compact* $(2m+1)$-*dimensional manifold $M$ of constant curvature $c > 0$ verifies*

$$\mathrm{Vol}(V) \geq \sum_{k=0}^{m} \frac{\binom{m}{k}^2}{\binom{2m}{2k}} r^{2k} c^k \,\mathrm{vol}(M)$$

*and for $m \neq 1$ equality never holds.*

***Proof*** Let $V$ be a vector field of constant length $r > 0$ on a Riemannian manifold $M$ of dimension $n$ and let us define $P$ as the restriction of $\nabla V$ to the subbundle $\mathscr{D}_V^{\perp}$. Since the extension of the metric $g$ to multi-vectors is given by $g(V_1 \wedge \ldots \wedge V_k, W_1 \wedge \ldots \wedge W_k) = \det(g(V_i, W_j))$, in a local orthonormal frame $\{E_i\}_{i=1}^n$ we have

$$\sigma_k((\nabla V)^t \circ \nabla V) = \sum_{j_1 < \ldots < j_k \leq n} \|\nabla_{E_{j_1}} V \wedge \ldots \wedge \nabla_{E_{j_k}} V\|^2 \tag{4.19}$$

and if, in particular, we choose $E_n = \frac{1}{r}V$ we obtain that

$$\sigma_k((\nabla V)^t \circ \nabla V) \geq \sigma_k(P^t \circ P) = \sum_{j_1 < \ldots < j_k < n} \|P(E_{j_1}) \wedge \ldots \wedge P(E_{j_k})\|^2 \tag{4.20}$$

with equality if and only if $\nabla_V V = 0$. On the other hand

$$\|P(E_{j_1}) \wedge \ldots \wedge P(E_{j_k})\|^2 = \sum_{i_1<\ldots<i_k} g(P(E_{j_1}) \wedge \ldots \wedge P(E_{j_k}), E_{i_1} \wedge \ldots \wedge E_{j_k})^2$$

$$= \sum_{i_1<\ldots<i_k} \det(P_{i_1,\ldots,i_k}^{j_1,\ldots,j_k})^2 \tag{4.21}$$

where $\det(P_{i_1,\ldots,i_k}^{j_1,\ldots,j_k})$ is the corresponding minor of order $k$.

Let's denote by $\{\mu_1, \mu_2 = \overline{\mu}_1, \ldots, \mu_{2s-1}, \mu_{2s} = \overline{\mu}_{2s-1}, \mu_{2s+1}, \ldots \mu_{n-1}\}$ the roots of the characteristic polynomial of $P$ that we assume to be complex functions from 1 to $2s$ and let's choose the local frame such that the matrix of $P$ has a canonical Jordan blocks decomposition. Then by (4.20), (4.21) and (4.16)

$$\sigma_k((\nabla V)^t \circ \nabla V) \geq \sum_{j_1<\ldots,<j_k} |\mu_{j_1}|^2 \cdots |\mu_{j_k}|^2 \geq \frac{1}{\binom{n-1}{k}} \Big( \sum_{j_1<\ldots,<j_k} |\mu_{j_1}| \cdots |\mu_{j_k}| \Big)^2.$$

The last inequality is strict unless all the real numbers $|\mu_j|$ are equal. Consequently

$$f_V^2 = \sum_{k=0}^{n} \sigma_k((\nabla V)^t \circ \nabla V) \geq \sum_{k=0}^{n-1} \frac{1}{\binom{n-1}{k}} \Big( \sum_{j_1<\ldots,<j_k} |\mu_{j_1}| \cdots |\mu_{j_k}| \Big)^2.$$

By hypothesis, the dimension of $M$ is $n = 2m + 1$ and we can apply Lemma 4.3 to obtain

$$f_V \geq \sum_{k=0}^{m} \frac{\binom{m}{k}}{\binom{2m}{2k}} \sum_{j_1<\ldots<j_{2k}} |\mu_{j_1}| \cdots |\mu_{j_{2k}}|$$

and using now (4.15)

$$f_V \geq \sum_{k=0}^{m} \frac{\binom{m}{k}}{\binom{2m}{2k}} \sigma_{2k}(P). \tag{4.22}$$

F. Brito, R. Langevin and H. Rosenberg showed in [20] that for unit vector fields on a manifold of constant curvature the integrals of the functions $\sigma_k(P)$ do not depend on the particular $V$ and then in our case the integrals only will depend on the length $r$. In particular, using Theorem 1.1 of [20] we get the equality

$$\int_M \sigma_{2k}(P) dv_g = \binom{m}{k} r^{2k} c^k \operatorname{vol}(M)$$

which combined with (4.22) completes the proof of the lower bound.

Let's now assume that for a vector field $V$ the equality holds, as we have pointed out during the proof, this would imply that $\nabla_V V = 0$ and that $|\mu_i| = |\mu_j|$ for all the eigenvalues of $P$. Moreover, Eq. (4.17) implies that except for $m = 1$ the equality in (4.15) for all the choices of indexes $j_1 < \ldots < j_{2k}$ would imply that all the eigenvalues of $P$ must be equal to a positive real number $\mu$ and $P = \mu \mathrm{Id}$. In other words the distribution $\mathscr{D}_V$ would be geodesic and the orthogonal distribution $\mathscr{D}_V^\perp$ would be integrable and with umbilical leaves.

F. G. B. Brito and P. Walczak in [18] have shown that if a complete manifold of non negative Ricci curvature admits an integrable distribution of co-dimension 1 such that the orthogonal distribution is geodesic then the manifold must be locally a Riemannian product. As a consequence, if there exists a vector field $V$ of constant length satisfying the equality, the sectional curvature of any plane generated by $V$ and a vector field in $\mathscr{D}_V^\perp$ should be zero which is in contradiction with the hypothesis of $M$ being of constant curvature $c > 0$. Then, for $m \neq 1$ the equality never holds. □

*Remark* The corresponding result for manifolds of constant negative curvature is also true. More precisely, the volume of a vector field $V$ of length $r > 0$ on a compact $(2m+1)$-dimensional manifold $M$ of constant curvature $c < 0$ verifies

$$\mathrm{Vol}(V) \geq \sum_{k=0}^{m} \frac{\binom{m}{k}^2}{\binom{2m}{2k}} r^{2k} |c|^k \mathrm{vol}(M).$$

In contrast with the case of the spherical space forms that have received much attention, the volume of vector fields on hyperbolic space forms is still a domain where very few results are known.

*Remark* Although, for the sake of completeness, we have stated Theorem 4.2 for every positive values $r$ and $c$ it's clear that the lower bound for vector fields of length $r$ on a space of constant curvature $c$ is the lower bound for unit vector fields on a space constant curvature $r^2|c|$ and it's enough to consider only unit vector fields, as in [21]. Equivalently, one can assume that the curvature is 1 without losing generality.

## 4.4 Asymptotic Behaviour of the Volume Functional

If we fix a unit vector field $V$ on a Riemannian manifold $M$ then (4.1) gives for all $r > 0$,

$$\mathrm{Vol}(rV) = \int_M \sqrt{1 + r^2\sigma_1((\nabla V)^t \nabla V) + \cdots + r^{2(n-1)}\sigma_{n-1}((\nabla V)^t \nabla V)}\, \mathrm{dv}_g$$

and if we assume that $M$ is compact, or more generally that $\mathrm{Vol}(V)$ is finite, it's easy to see, as we have pointed out in [11], that the asymptotic behaviour of the real function of one variable $r \to \mathrm{Vol}(rV)$, when $r$ tends to 0, is determined by the Energy of $V$. More precisely, if $V$ and $W$ are unit vector fields then, by elementary calculus, the function $h(r) = \mathrm{Vol}(rV) - \mathrm{Vol}(rW)$ verifies $h(0) = 0$, $\lim_{r\to 0} \frac{h(r)}{r} = 0$ and

$$\lim_{r\to 0} \frac{\mathrm{Vol}(rV) - \mathrm{Vol}(rW)}{r^2} = \int_M \|\nabla V\|^2 \mathrm{dv}_g - \int_M \|\nabla W\|^2 \mathrm{dv}_g. \tag{4.23}$$

The function $h(r)$ has at $r = 0$ an infinitesimal of order at least 2, the order is 2 if $E(V) \neq E(W)$.

When $r$ tends to infinity, the function $\mathrm{Vol}(rV)$ is dominated by the term of highest degree on $r$. The functions $\frac{\mathrm{Vol}(rV)}{r^k}$, for $k = 0, \ldots, n-2$, tend to infinity and so, the function $\mathrm{Vol}(rV)$ is an infinitely large quantity of order $(n-1)$, since

$$\lim_{r\to\infty} \frac{\mathrm{Vol}(rV)}{r^{n-1}} = \int_M \sqrt{\sigma_{n-1}((\nabla V)^t \nabla V)} \mathrm{dv}_g. \tag{4.24}$$

This fact led us to define in [11] the *Twisting functional* precisely as

$$T(V) = \int_M \sqrt{\sigma_{n-1}((\nabla V)^t \nabla V)} \mathrm{dv}_g,$$

which determines the asymptotic behaviour of the function $r \to \mathrm{Vol}(rV)$ when $r$ tends to infinity.

The following result is a direct consequence of (4.23) and (4.24).

**Proposition 4.3** *Let $V$, $W$ be unit vector fields with finite volume. If $E(V) > E(W)$ then there exits $r(V, W) > 0$ such that* $\mathrm{Vol}(rV) > \mathrm{Vol}(W)$ *for all $r < r(V, W)$. If $T(V) > T(W)$ then there exits $\overline{r}(V, W) > 0$ such that* $\mathrm{Vol}(rV) > \mathrm{Vol}(rW)$ *for all $r > \overline{r}(V, W)$.*

Now we consider the particular case of $M = S^{2m+1}$, or any complete manifold of curvature 1.

**Corollary 4.1** *The Hopf vector fields on a $(2m+1)$-dimensional manifold $M$ of curvature* 1 *minimise the twisting among unit vector fields. Moreover, if a unit vector field $V$ verifies $T(V) > T(H)$ there is $\overline{r}_V > 0$ such that* $\mathrm{Vol}(rV) > \mathrm{Vol}(rH)$ *for all $r > \overline{r}_V$.*

***Proof*** Since $\nabla H = J$ on $\mathscr{D}_{H^\perp}$ and $\nabla_H H = 0$, it's evident that $T(H) = \mathrm{vol}(M)$. Using the same notation as in Theorem 4.2,

$$\sigma_{2m}((\nabla V)^t \nabla V) \geq \sigma_{2m}(P^t P) = \det(P)^2 = \sigma_{2m}(P)^2$$

and consequently

$$T(V) \geq \int_M \sigma_{2m}(P)\mathrm{dv}_g.$$

As in Theorem 4.2, the result holds now from Theorem 1.1 of [20]; the second assertion follows from Proposition 4.1 with $W = H$ and with the notation $\overline{r}_V = \overline{r}(V, H)$. □

**Corollary 4.2** *Let $M$ be a $(2m+1)$-dimensional manifold of curvature 1 different from $S^{2m+1}$. The Hopf vector fields are the only minimisers of the Energy. Consequently for any unit vector field $V$ which is not a Hopf vector field there is $r_V > 0$ such that $\mathrm{Vol}(rV) > \mathrm{Vol}(rH)$ for all $r < r_V$.*

***Proof*** The first assertion is an immediate consequence of Proposition 3.7 about the first eigenvalue of the rough Laplacian that for a unit vector field implies that

$$\int_M \|\nabla V\|^2 \mathrm{dv}_g \geq 2m \int_M \|V\|^2 \mathrm{dv}_g = 2m\mathrm{vol}(M) = \int_M \|\nabla H\|^2 \mathrm{dv}_g$$

with equality only if $V$ is a Killing vector field. The second assertion follows from the proposition above. □

The first part of the Corollary above was shown by D. Perrone in [82] with a different method.

In the case of $M = S^{2m+1}$ the first eigenvalue of the rough Laplacian is 1 (Proposition 3.3) and then the estimate it gives for unit vector fields is

$$\int_M \|\nabla V\|^2 \mathrm{dv}_g \geq \int_M \|V\|^2 \mathrm{dv}_g = \mathrm{vol}(M)$$

and the equality never holds since there are no unit vector fields among the eigenvectors corresponding to the first eigenvalue. But this bound shows up to be irrelevant for $m \neq 1$ because of the improvement by F. Brito and P.G. Walczak in [18].

**Proposition 4.4** *On the sphere $S^{2m+1}$ with $m > 1$, for every unit vector field $V$*

$$\int_{S^{2m+1}} \|\nabla V\|^2 \mathrm{dv}_g > (1 + \frac{1}{2m-1})\mathrm{vol}(S^{2m+1}).$$

***Proof*** With the same notation used in Theorem 4.1 we have that for a unit vector field $V$

$$\|\nabla V\|^2 = \sigma_1((\nabla V)^t \circ \nabla V) = \sigma_1(P^t \circ P) + \|\nabla_V V\|^2 \geq \sigma_1(P^t \circ P),$$

with equality if and only if $\nabla_V V = 0$. From (4.6) we have that $(2m-1)\sigma_1(P^t \circ P) \geq 2\sigma_2(P)$ with equality if and only if

$$\sum_{i<j}\left((P_{ii} - P_{jj})^2 + (P_{ij} + P_{ji})^2\right) = 0 \quad \text{and} \quad \sum_{i \neq j}^{k} P_{ij}^2 = 0.$$

The lower bound follows from (4.3) and the fact that it can't be attained by any smooth unit vector field is obtained as in Theorem 4.2, because also here the equality would imply $\nabla_V V = 0$ and $P = \mu \mathrm{Id}$ with $\mu = P_{ii}$, which is impossible. □

## 4.5 Notes

### 4.5.1 *Unit Vector Fields on the Two-Dimensional Torus*

A two-dimensional compact manifold admitting a unit vector field must be topologically a torus. In [65] the volume of a flow is defined to be the volume of a unit vector field that generates the flow.The following result was shown: For any homotopy class of flows on a two-dimensional torus with any given Riemannian metric, there is a smooth flow of smallest area, unique up to rotation by a constant angle.

**Open Problem** The author has used the method of calibrations. It would be interesting to try a new proof by using the eigenvalues of the rough Laplacian as in Theorem 4.1 and to determine the minimising vector fields.

### 4.5.2 *Lower Bound of the Volume of Unit Vector Fields on Hypersurfaces of* $\mathbf{R}^{n+1}$

A. G. Reznikov studied in [84] the volume of unit vector fields $V$ on a compact orientable submanifold $M$ of $\mathbf{R}^{n+1}$ of co-dimension 1 and found a lower bound in terms of the Gauss map. The bound, obtained by comparing the endomorphisms field $\nabla V$ with the shape operator $S$, in conveniently chosen local frames, is

$$\mathrm{Vol}(V) \geq \mathrm{vol}(M) + \frac{|\beta(M)|}{\kappa}\mathrm{vol}(S^n),$$

where $\beta(M)$ is the degree of the Gauss map and $\kappa = \sup_{x \in M} \|S_x\|$.

It's noteworthy that for $M = S^{2m+1}$ the inequality only gives

$$\mathrm{Vol}(V) \geq 2\mathrm{vol}(S^{2m+1})$$

which apart from the special case $m = 1$ is a result weaker than the one in [52].

### 4.5.3 *Almost Hermitian Structures on $S^6$ That Minimise the Volume*

An almost Hermitian structure on a Riemannian manifold $(M, g)$ is a section of the fiber bundle $\mathscr{H}(M)$ having as fiber over $x \in M$ the set all the complex structures of $(T_x M, g_x)$. The only spheres admitting almost Hermitian structures are $S^2$ and $S^6$ but since $\mathscr{H}(S^2)$ consists in two copies of $S^2$, only for the 6-dimension case the problem of finding the section of minimum volume can be posed. This was done by E. Calabi and H. Gluck in [28] where they showed that the minimisers are exactly the Cayley structures and no others. The standard Cayley structure is defined by identifying $S^6$ with the unit purely imaginary Cayley numbers and then for $x \in S^6$ one can define the almost complex structure of $T_x S^6$ as right multiplication by $x$.

For the proof, the authors use that $\mathscr{H}(S^6) = O(8)/U(4)$ and show that one of the connected components of this space $SO(8)/U(4)$ is isometric to $\mathrm{Gr}_2(\mathbf{R}^8)$, the Grassmanian of oriented two planes of $\mathbf{R}^8$. Then they show that the image of a Cayley structure is a sub-Grassmanian of type $\mathrm{Gr}_1(\mathbf{R}^7)$, of lines of $\mathbf{R}^7$. The conclusion is then obtained directly as a consequence of the work [53] on calibrated geometries in Grassman manifolds.

### 4.5.4 *Minimisers of Functionals Related with the Energy*

In [16], F. G. B. Brito has defined a new functional, the *corrected energy*, by the expression

$$CE(V) = E(V) + \frac{1}{2}\int_M (n-1)(n-3)\|h_{V^\perp}\|^2 \mathrm{dv}_g,$$

where $n$ is the dimension of $M$ and $h_{V^\perp}$ is the mean curvature vector field of the distribution $\mathscr{D}_V^\perp$; he shows that the Hopf vector fields minimise this new functional.

The definition of the corrected energy functional has been extended from vector fields to distributions by P. M. Chacón and A.M. Naveira in [30], in particular if $(M, g)$ is a $n$-dimensional manifold and we consider a section $\sigma : (M, g) \to (G_q(M), g^S)$ of its Grassmanian of $q$-planes, i. e. a distribution of dimension $q$, with energy $E(\sigma)$ the authors define its corrected energy as

$$CE(\sigma) = E(\sigma) + \frac{1}{2}\int_M \Big((n-q)(n-q-2)\|h_{\sigma^\perp}\|^2 + q^2\|h_\sigma\|^2\Big)\mathrm{dv}_g,$$

where $h_\sigma, h_{\sigma^\perp}$ represent the mean curvature vector fields of $\sigma$ and $\sigma^\perp$, respectively.

With this definition, the authors showed that among the 3-dimensional integrable distributions on $S^{4m+3}$, those of minimum energy are the distributions defined by the subspaces tangent to the fibers of the quaternionic Hopf fibration $\pi : S^{4m+3} \to \mathbf{H}P^m$.

In [45] we have studied the infimum of the energy of unit vector fields on the three-dimensional Berger sphere $(S^3, g_\mu)$ with the same method referred to in the proof of Proposition 4.2 for the volume functional. In particular we have shown that if $\mu < 1$ then the energy of a unit vector field $V$ verifies

$$E(V) \geq E(H_\mu) = \Big(\frac{3}{2} + \mu\Big)\mathrm{vol}(S^3, g_\mu)$$

with equality if and only if $V = \pm H^\mu$. If $\mu > 1$, for all unit vector fields $X$ in the 2-dimensional space generated by $\{E_1, E_2\}$ we have

$$E(X) = \Big(\frac{3}{2} + \mu + 2\frac{1-\mu}{\mu}\Big)\mathrm{vol}(S^3, g_\mu) < E(H_\mu).$$

As in the case of the volume, we don't know if these vector fields are the minimisers.

F. G. B. Brito, I. Gonçalves and A. V. Nicoli have found, in [25], a lower bound of the energy of unit vector fields on hypersurfaces of the Euclidean space, following the method of [84] for the volume described in Sect. 4.5.2. More precisely they have shown that if $M$ is a closed oriented odd-dimensional hypersurface of $\mathbf{R}^{2m+2}$ different from the sphere, then for every unit vector field $V$

$$E(V) \geq \frac{m|\beta(M)|}{(2m-1)\overline{\kappa}(M)}\mathrm{vol}(S^{2m+1}) + \frac{2m+1}{2}\mathrm{vol}(M),$$

where $\overline{\kappa}(M)$ is a constant depending on the shape operator of $M$.

Another functional, similar to the energy, has been considered in [25] where the authors have shown that for any compact oriented Riemannian manifold $M$ of dimension $2m+1$ and for any unit vector field $V$

$$\int_M \sigma_m((\nabla V)^t \nabla V))\mathrm{dv}_g \geq \binom{2n}{n}\int_M |\sigma_{2m}(P)|\mathrm{dv}_g$$

and that the Hopf vector fields of odd-dimensional spheres minimise also this integral.

# Chapter 5
# Vector Fields of Constant Length on Punctured Spheres

With the term punctured sphere we will refer to an open subset of the sphere whose complementary is a finite set.

In the first section, we study vector fields of constant length tangent to the radial geodesics, issuing from a given point $a \in S^n$ they are smooth on $S^n \setminus \{a, -a\}$. Radial vector fields are $r$-minimal and, if $n = 2m + 1$, their volume is the lower bound of Theorem 4.2.

Another family of vector fields of constant length is studied in the second section. They are obtained by parallel transport of a given vector, tangent to the sphere at a point $a \in S^n$, along the radial geodesics issuing from $a$ and they are smooth on $S^n \setminus \{-a\}$. Parallel transport vector fields are $r$-minimal if and only if $r = 1$, except for the case $n = 2$ when they are $r$-minimal for all $r$. As far as we know this is the only example of unit minimal vector field such that the vector fields obtained by scaling are not $r$-minimal. Moreover, we will see that the volume of these vector fields is the limit of the volume of smooth vector fields.

Vector fields defined on punctured spheres, also referred as singular vector fields on the sphere, have an important influence on the main open problem of determining the infimum of the volume of smooth vector fields of constant length on the odd-dimensional spheres. We devote Sect. 5.3 to the state of the art in the problem.

If we deal with vector fields defined on punctured spheres, it's natural to extend the problem of finding the volume minimising vector fields also to even-dimensional spheres. In section four we will see that the parallel transport vector fields are the area minimising vector fields of constant length on $S^2$, among those who are as regular as possible.



O. Gil-Medrano, *The Volume of Vector Fields on Riemannian Manifolds*,
Lecture Notes in Mathematics 2336, https://doi.org/10.1007/978-3-031-36857-8_5

## 5.1 The Radial Vector Fields

Radial vector fields on the sphere have played an important role in previous chapters in its condition of eigenvectors of the first eigenvalue of the rough Laplacian acting on vector fields $\lambda_1^* = 1$ (see Proposition 3.3). Indeed for $a \in \mathbf{R}^{n+1}$, $a \neq 0$, the vector field on $S^n$ defined as $\overline{R}^a = -\operatorname{grad} f_a$, where $f_a = g(a, N)$, is tangent to the great circles passing through $a/\|a\|$, more precisely

**Lemma 5.1** *For every $a \in \mathbf{R}^{n+1}$ with $\|a\| = 1$ the integral curve of $\overline{R}^a$ passing through $p \neq \pm a$ is the great circle from $a$ to $-a$ parametrised as follows*

$$c_p(t) = -\tanh(t)a + \frac{1}{\cosh(t)} \frac{(p - f_a(p)a)}{\sqrt{1 - f_a(p)^2}}$$

*and the curve $c_p$ passes trough $p$ for $t_p = \tanh^{-1}(-f_a(p))$.*

***Proof*** Since $\overline{R}^a = f_a N - a$ and $f_a(c_p(t)) = -\tanh(t)$, we obtain that

$$\overline{R}^a(c_p(t)) = -(1 - \tanh^2(t))a + \frac{\sinh(t)}{\cosh^2(t)} \frac{(p - f_a(p)a)}{\sqrt{1 - f_a(p)^2}}.$$

Then the equality $\overline{R}^a(c_p(t)) = c_p'(t)$ is an easy consequence. □

**Definition 5.1** A radial vector field of constant length $r > 0$ of $S^n$ will be a vector field of the form

$$R^a = \frac{-r}{\|\operatorname{grad} f_a)\|} \operatorname{grad} f_a = \frac{-r}{\sqrt{1 - f_a^2}}(a - f_a N),$$

for $a \in \mathbf{R}^{n+1}$ with $\|a\| = 1$. $R^a$ is tangent to the radial geodesics issuing from $a$ parametrised proportionally to their arc length and is defined in $S^n \setminus \{a, -a\}$.

**Proposition 5.1** *A radial vector field $R$ of constant length $r > 0$ is $r$-minimal.*

***Proof*** Let $R$ be a radial vector field, $R = R^a$ with $\|a\| = 1$, and let $V$ be any vector field on $S^n \setminus \{a, -a\}$. If we denote $\overline{\nabla}$ the covariant derivative of the Euclidean metric on $\mathbf{R}^{n+1}$ then, $\overline{\nabla}_V \overline{R} = V(f_a)N + f_a \overline{\nabla}_V N$, where $\overline{R} = \overline{R}^a = -\operatorname{grad} f_a$. Consequently, $\nabla_V \overline{R} = f_a V$ and we have

$$\nabla_V R = \frac{r f_a}{\sqrt{1 - f_a^2}} V + \frac{f_a\, g(a, V)}{1 - f_a^2} R. \tag{5.1}$$

In particular if $V \in \mathscr{D}_R^{\perp}$, which is equivalent to $V(f_a) = g(a, V) = 0$, then

$$\nabla_V R = \frac{r f_a}{\sqrt{1 - f_a^2}} V.$$

Moreover, as we know in advance by the definition, $\nabla_R R = 0$. Now

$$L_R = \mathrm{Id} + (\nabla R)^t \circ \nabla R$$

$$= \Big(1 + \frac{r^2 f_a^2}{1 - f_a^2}\Big)\mathrm{Id} \quad \text{on } \mathscr{D}_R^{\perp} \quad \text{and} \quad L_R = \mathrm{Id} \quad \text{on} \quad \mathscr{D}_R,$$

$$f_R = \Big(1 + \frac{r^2 f_a^2}{1 - f_a^2}\Big)^{(n-1)/2}, \tag{5.2}$$

$$K_R = f_R \nabla R \circ L_R^{-1} = \Big(1 + \frac{r^2 f_a^2}{1 - f_a^2}\Big)^{(n-3)/2} \frac{r f_a}{\sqrt{1 - f_a^2}} \mathrm{Id} \quad \text{on} \quad \mathscr{D}_R^{\perp} \quad \text{and}$$

$$K_R = 0 \quad \text{on} \quad \mathscr{D}_R.$$

To compute the divergence of $K_R$ we consider a local orthonormal frame $\{E_i\}_{i=1}^n$ with $E_n = R/r$. Since $\nabla_R R = 0$ and $K_R(R) = 0$, for $i = 1, \ldots, n-1$,

$$K_R(\nabla_{E_i} E_i) = K_R(\nabla_{E_i} E_i - g(\nabla_{E_i} E_i, R)R).$$

By (5.1),

$$g(\nabla_{E_i} E_i, R) = -g(E_i, \nabla_{E_i} R) = -\frac{r f_a}{\sqrt{1 - f_a^2}}$$

and then, by (5.2),

$$\nabla^* K_R = -\sum_{i=1}^{n-1} \Big\{\nabla_{E_i} K_R(E_i) - K_R(\nabla_{E_i} E_i + \frac{f_a}{r\sqrt{1 - f_a^2}} R)\Big\}$$

$$= \Big(1 + \frac{r^2 f_a^2}{1 - f_a^2}\Big)^{(n-3)/2} \frac{f_a^2}{1 - f_a^2} R, \tag{5.3}$$

because

$$E_i\Big(\Big(1 + \frac{r^2 f_a^2}{1 - f_a^2}\Big)^{(n-3)/2} \frac{r f_a}{\sqrt{1 - f_a^2}}\Big) = 0.$$

Thus, $R$ is $r$-minimal from Definition 2.2. □

For odd-dimensional spheres the volume and the energy of these vector fields are exactly the lower bounds of Theorem 4.2 and Proposition 4.4, respectively.

**Proposition 5.2** *The volume of a radial vector field of length $r$ on $S^{2m+1}$ verifies $\mathrm{Vol}(R) = c(m;r)\mathrm{vol}(S^{2m+1})$ with*

$$c(m;r) = \sum_{k=0}^{m} \frac{\binom{m}{k}^2}{\binom{2m}{2k}} r^{2k}.$$

*For a unit radial vector field*

$$\int_{S^{2m+1}} \|\nabla R\|^2 \mathrm{dv}_g = (1 + \frac{1}{2m-1})\mathrm{vol}(S^{2m+1}).$$

***Proof*** We choose for $S^{2m+1}$ the usual spherical coordinates $\{\varphi_1, \ldots, \varphi_{2m}, t\}$ modulo the identification of an orthonormal basis of the form $\{e_1, \ldots, e_{2m+1}, a\}$ with the canonical one. With this choice, $f_a = \cos t$, $f_R$ is a function that only depends on $t$ and

$$\begin{aligned}
\mathrm{Vol}(R) &= \mathrm{vol}(S^{2m}) \int_0^{\pi} \left(1 + \frac{r^2 \cos^2 t}{1 - \cos^2 t}\right)^m (\sin t)^{2m}\, dt \\
&= \mathrm{vol}(S^{2m}) \sum_{k=0}^{m} \binom{m}{k} r^{2k} \int_0^{\pi} (\cos t)^{2k} (\sin t)^{2m-2k}\, dt \\
&= 2\mathrm{vol}(S^{2m}) \sum_{k=0}^{m} \binom{m}{k} r^{2k} \int_0^{\infty} (1+u^2)^{-(2m+1)} (1-u^2)^{2k} (2u)^{2(m-k)}\, du \\
&= \frac{\mathrm{vol}(S^{2m})}{\Gamma(1+m)} \sum_{k=0}^{m} \binom{m}{k} r^{2k} \Gamma(\frac{1}{2} + m - k)\Gamma(\frac{1}{2} + k).
\end{aligned} \tag{5.4}$$

From the well know expression of the volume of the spheres and the $\Gamma$ function

$$\mathrm{vol}(S^n) = \frac{2\pi^{(n+1)/2}}{\Gamma(\frac{n+1}{2})} \qquad \Gamma(\frac{1}{2} + k) = \frac{\pi^{\frac{1}{2}}(2k)!}{4^k k!} \tag{5.5}$$

we obtain

$$\begin{aligned}
\mathrm{vol}(S^{2m}) &= \frac{4^m}{\pi} \binom{2m}{m}^{-1} \mathrm{vol}(S^{2m+1}) \\
\Gamma(\frac{1}{2} + m - k)\Gamma(\frac{1}{2} + k) &= \frac{\pi(2m-2k)!(2k)!}{4^m (m-k)!k!}.
\end{aligned} \tag{5.6}$$

These equalities jointly with 5.4 give

$$\mathrm{Vol}(R) = \binom{2m}{m}^{-1} \sum_{k=0}^{m} \binom{2k}{k} \binom{2m-2k}{m-k} r^{2k} \, \mathrm{vol}(S^{2m+1}) = c(m;r)\mathrm{vol}(S^{2m+1}).$$

For the second part, assume that $r = 1$ then

$$\begin{aligned}
\int_{S^{2m+1}} \|\nabla R\|^2 \mathrm{dv}_g &= 2m \int_{S^{2m+1}} \frac{f_a^2}{1-f_a^2} \mathrm{dv}_g \\
&= 2m \, \mathrm{vol}(S^{2m}) \int_0^{\pi} \cos^2 t (\sin t)^{2m-2} \, dt \\
&= 2m \, \mathrm{vol}(S^{2m}) \frac{\Gamma(\frac{1}{2}+m-1)\Gamma(\frac{1}{2}+1)}{\Gamma(1+m)}
\end{aligned}$$

and the result is obtained by (5.5) and (5.6). □

*Remark* In our joint work with Borrelli and Brito [14] and [15] we have shown that the infimum of the energy of unit smooth vector fields on $S^{2m+1}$ (for $m > 1$) is the energy of the radial vector fields. To do so we have constructed for each radial unit vector field $R = R^a$ a family of unit smooth vector fields $\{R_\varepsilon \, ; \, 1 \geq \varepsilon > 0\}$ which are equal to $R$ outside a neighbourhood of a geodesic segment joining the points $a$ and $-a$ and with energy converging to the energy of $R$ as $\varepsilon$ tends to 0. This result is in contrast with the case of the other $(2m+1)$-dimensional manifolds of curvature 1 different from $S^{2m+1}$ (see Corollary 4.2) for which the Hopf vector fields are the only minimisers of the energy.

## 5.2 Parallel Transport Vector Fields

Another family of interesting vector fields are those obtained by parallel transport along radial geodesics issuing from of a point $p \in S^n$ of a given vector $v \neq 0$ tangent to $S^n$ at this point. These vector fields, of constant length by construction, are defined in $S^n \setminus \{-p\}$. They where used in the context of the volume of vector fields problems by the first time by S. Pedersen in [78] where she showed that for $n = 2m+1$ there is a curve of smooth unit vector fields defined on the entire sphere whose volumes converge to the volume of the parallel transport vector fields; the author also showed that for $m > 1$ their volume is strictly lower than the volume of the Hopf vector fields and conjectured that this value could be the infimum of the volume of smooth unit vector fields on the sphere. This conjecture is open, but wc are going to show that it can't be extended to vector fields of constant length $r \neq 1$.

**Definition 5.2** To each orthonormal 2-frame $\{a, b\}$ of $\mathbf{R}^{n+1}$ and $r > 0$ we can associate the smooth vector field $P^{a,b}$ defined in $S^n \setminus \{-a\}$ as follows: $P^{a,b}(p)$

is the parallel transport of $rb$ along the unique geodesic from $a$ to $p$ contained in $S^n \setminus \{-a\}$. Any vector field constructed by this method will be called a parallel transport vector field of length $r$.

**Lemma 5.2** *Parallel transport vector fields of length $r$ are of the form*

$$P = r\left(b - \frac{f_b}{1+f_a}(a+N)\right) \tag{5.7}$$

*with $\{a, b\}$ being an orthonormal 2-frame of $\mathbf{R}^{n+1}$ and $P$ defined on $S^n \setminus \{-a\}$.*

***Proof*** We only need to show that $\nabla_R P = 0$ for $R = R^a$ the unit radial vector field. Let $V$ be any vector field then

$$\overline{\nabla}_V P = -V(\frac{rf_b}{1+f_a})(a+N) - \frac{rf_b}{1+f_a}V$$

and consequently

$$\begin{aligned}
\nabla_V P &= -V(\frac{rf_b}{1+f_a})(a - f_a N) - \frac{rf_b}{1+f_a}V \\
&= -r\frac{(1+f_a)g(b,V) - f_b g(a,V)}{(1+f_a)^2}(a - f_a N) - \frac{rf_b}{1+f_a}V \\
&= -\frac{g(P,V)}{1+f_a}(a - f_a N) - \frac{rf_b}{1+f_a}V.
\end{aligned} \tag{5.8}$$

In particular, for $V = R$ since $g(P, R) = \frac{rf_b}{\sqrt{1-f_a^2}}$ we obtain $\nabla_R P = 0$. □

These vector fields for the case $r = 1$ are described in [78] as the submanifolds of the Stiefel manifold obtained by the action of $O(n)$ and this description leads to the minimality of their images in this homogeneous manifold. As we have pointed out in previous chapters, the Sasaki metric on the bundle of vectors of length $r \neq 1$ is not the homogeneous one; in fact we are going to show that parallel transport vector fields are not minimal in general, except for the case of the 2-spheres. The proof below follows the scheme we have used in [47] for checking directly that for $r = 1$ they are critical points of the volume functional.

**Proposition 5.3** *For $n \neq 2$, parallel transport vector fields of length $r$ are $r$-minimal if and only if $r = 1$. Parallel transport vector fields of length $r$ on $S^2$ are $r$-minimal for all $r$.*

***Proof*** Let $P$ be the parallel transport vector field of length $r$ corresponding to the orthonormal 2-frame $\{a, b\}$ of $\mathbf{R}^{n+1}$, then by (5.8)

$$\nabla_V P = -\frac{rf_b}{1+f_a}V \quad \text{if } V \in \mathcal{D}_P^{\perp} \quad \text{and} \quad \nabla_P P = \frac{r^2}{1+f_a}\overline{R}^a - \frac{rf_b}{1+f_a}P$$

Let's complete $\{a, b\}$ to an orthonormal basis $\{e_i\}_{i=1}^{n+1}$ of $\mathbf{R}^{n+1}$ such that $e_{n+1} = a$, $e_n = b$ and let's consider the orthonormal frame $\{E_i\}_{i=1}^{n}$ defined in $S^n \setminus \{-a\}$ by $E_i = P^{a,e_i}$; in particular $E_n = \frac{1}{r} P$.The matrix of $\nabla P$ in this frame is

$$\nabla P = \frac{r}{1+f_a} \begin{pmatrix} & & f_1 \\ & -f_b \mathrm{Id} & \vdots \\ & & f_{n-1} \\ 0 & \cdots \quad \cdots & 0 \end{pmatrix} \tag{5.9}$$

where we have represented for simplicity the functions $f_{e_i}$ by $f_i$ and used that $\sum_{j=1}^{n} f_j E_j = \overline{R}^a$. Then the endomorphism field $L_P$ is given by

$$L_P(E_i) = \sum_{j=1}^{n} g(L_P(E_i), E_j) E_j = \sum_{j=1}^{n} \Big(\delta_{ij} + g(\nabla P(E_i), \nabla P(E_j))\Big) E_j$$

$$= \Big(1 + \frac{r^2 f_b^2}{(1+f_a)^2}\Big) E_i - \frac{r^2 f_b}{(1+f_a)^2} f_i E_n \quad \text{for} \quad i < n \quad \text{and} \tag{5.10}$$

$$L_P(E_n) = -\sum_{j=1}^{n-1} \frac{r^2 f_b}{(1+f_a)^2} f_j E_j + \Big(1 + \frac{r^2(1 - f_a^2 - f_b^2)}{(1+f_a)^2}\Big) E_n. \tag{5.11}$$

It is easy to see that for a $n \times n$ matrix $A$ of the form

$$A = \begin{pmatrix} \lambda \mathrm{Id} & \beta^t \\ \beta & \varepsilon \end{pmatrix} \tag{5.12}$$

with $\lambda, \varepsilon \in \mathbf{R}$ and $\beta \in \mathbf{R}^{n-1}$ we have $\det A = \lambda^{n-2}(\lambda\varepsilon - \|\beta\|^2)$ and if $\det A \neq 0$ then

$$A^{-1} = \frac{1}{\lambda(\lambda\varepsilon - \|\beta\|^2)} \begin{pmatrix} (\lambda\varepsilon - \|\beta\|^2)\mathrm{Id} + \beta^t \beta & -\lambda\beta^t \\ -\lambda\beta & \lambda^2 \end{pmatrix} \tag{5.13}$$

Using these formulas for the matrix of the endomorphism field $L_P$ in the chosen frame, whose entries are given by (5.10) and (5.11), and the notation

$$\psi_r = (1+f_a)^2 + r^2 f_b^2 \qquad \varphi_r = (1+r^2) + (1-r^2) f_a$$

we obtain

$$f_P = \left(1+\frac{r^2 f_b^2}{(1+f_a)^2}\right)^{(n-2)/2}\left(1+\frac{r^2(1-f_a^2)}{(1+f_a)^2}\right)^{1/2} = \frac{\psi_r^{(n-2)/2}\varphi_r^{1/2}}{(1+f_a)^{(2n-3)/2}}, \quad (5.14)$$

$$L_P^{-1}(E_i) = \frac{(1+f_a)^2}{\psi_r}E_i + \frac{r^4 f_b^2 f_i}{\varphi_r\psi_r(1+f_a)}\sum_{j=1}^{n-1} f_j E_j + \frac{r^2 f_b f_i}{\varphi_r(1+f_a)}E_n,$$

for $i < n$, and

$$L_P^{-1}(E_n) = \frac{r^2 f_b}{\varphi_r(1+f_a)}\sum_{j=1}^{n-1} f_j E_j + \frac{\psi_r}{\varphi_r(1+f_a)}E_n.$$

It will be more convenient to write the expression of $L_P^{-1}$ in a different form, using that $\sum_{j=1}^{n-1} f_j E_j = \overline{R}^a - f_b E_n$,

$$L_P^{-1}(E_i) = \frac{(1+f_a)^2}{\psi_r}E_i + \frac{r^4 f_b^2 f_i}{\varphi_r\psi_r(1+f_a)}\overline{R}^a + \frac{r^2 f_b f_i(1+f_a)}{\varphi_r\psi_r}E_n, \quad (5.15)$$

for $i < n$, and

$$L_P^{-1}(E_n) = \frac{r^2 f_b}{\varphi_r(1+f_a)}\overline{R}^a + \frac{(1+f_a)}{\varphi_r}E_n. \quad (5.16)$$

Finally, from (5.15), (5.16) and (5.9) and taking into account that $P$ is parallel along the radial geodesics, the endomorphism field $K_P = f_P\nabla P \circ L_P^{-1}$ is given by

$$K_P(E_i) = -\frac{r f_P(1+f_a) f_b}{\psi_r}\left(E_i - \frac{r^2 f_i}{(1+f_a)\varphi_r}(\overline{R}^a - f_b E_n)\right) \quad \text{for} \quad i < n \quad \text{and}$$

$$K_P(E_n) = \frac{r f_P}{\varphi_r}(\overline{R}^a - f_b E_n). \quad (5.17)$$

By (5.8) and (5.17) it's easy to see that

$$\sum_{i=1}^{n} \nabla_{E_i} E_i = \frac{n-1}{(1+f_a)}\overline{R}^a \quad \text{and} \quad K_P\left(\sum_{i=1}^{n} \nabla_{E_i} E_i\right) = 0. \quad (5.18)$$

Let's denote $K_i^j$ the components of $K_P$, we obtain by straightforward computation that

$$\sum_{i=1}^{n} f_i K_i^j = r^2\Big(-1 + f_a^2 + \sum_{i=1}^{n} f_i^2\Big) = 0, \tag{5.19}$$

$$\sum_{k=1}^{n} f_k K_j^k = -\frac{r f_P (1+f_a) f_b f_j}{\psi_r}\Big(1 - \frac{r^2(1 - f_a^2 - f_b^2)}{(1+f_a)\varphi_r}\Big) \quad \text{and}$$

$$\sum_{k=1}^{n} K_k^k = -\frac{r f_P (1+f_a) f_b}{\psi_r}\Big((n-1) - \frac{r^2(1 - f_a^2 - f_b^2)}{(1+f_a)\varphi_r}\Big).$$

Then for $j < n$

$$\begin{aligned} g(\nabla^* K_P, E_j) &= -\sum_{i=1}^{n} E_i(K_i^j) - \sum_{i,k=1}^{n} K_i^k g(\nabla_{E_i} E_k, E_j) \\ &= -\sum_{i=1}^{n} E_i(K_i^j) + \frac{1}{1+f_a}\sum_{i,k=1}^{n} K_i^k(\delta_{ij} f_k - \delta_{ik} f_j)) \\ &= -\sum_{i=1}^{n} E_i(K_i^j) + \frac{1}{1+f_a}\Big(\sum_{k=1}^{n} f_k K_j^k - f_j \sum_{k=1}^{n} K_k^k\Big) \\ &= -\sum_{i=1}^{n} E_i(K_i^j) + (n-2)\frac{r f_P f_b}{\psi_r}. \end{aligned} \tag{5.20}$$

Now it is convenient to write $E_i = e_i - \frac{f_i}{f_a} a + \frac{f_i}{f_a(1+f_a)}\overline{R}^a$ on the open and dense subset $U \subset S^n \setminus \{-a\}$ where $f_a \neq 0$ and taking into account that by (5.19)

$$\overline{R}^a\big(\sum_{i=1}^{n} f_i K_i^j\big) = 0 \quad \text{and} \quad \sum_{i=1}^{n} f_i\, \overline{R}^a(K_i^j) = 0$$

we can write on $U$

$$\sum_{i=1}^{n} E_i(K_i^j) = \sum_{i=1}^{n} e_i(K_i^j) - \frac{1}{f_a}\sum_{i=1}^{n} f_i\, a(K_i^j),$$

where the derivation of the functions $K_i^j$ with respect to the vector fields $\{e_i, a\}$ of $\mathbf{R}^{n+1}$ is well defined since the functions are defined in an open set of $\mathbf{R}^{n+1}$

containing $S^n \setminus \{-a\}$. It's easy to see that

$$\sum_{i=1}^{n} e_i(K_i^j) = (n-1)\frac{2r^3 f_P f_b f_j}{\varphi_r \psi_r} \tag{5.21}$$

and that by (5.19)

$$\sum_{i=1}^{n} f_i \; a(K_i^j) = a(\sum_{i=1}^{n} f_i \; K_i^j) = 2 f_a \frac{r^3 f_P f_b f_j}{\varphi_r \psi_r}. \tag{5.22}$$

Therefore, as equalities (5.20), (5.21) and (5.22) give that for $j < n$

$$g(\nabla^* K_P, E_j) = (n-2)(1-r^2)\frac{r f_P f_b f_j}{\varphi_r \psi_r},$$

the vector field $P$ is $r$-minimal if and only if $n = 2$ or $r = 1$. □

The 2-dimensional part of previous result was shown in [12] by a more geometrical method based in the isometry of $(T^r S^2, g^S)$ with a projective space with a Berger metric which will be described in the next section.

**Proposition 5.4** *The volume of a parallel transport vector field of length $r$ on $S^{2m+1}$ verifies* $\mathrm{Vol}(P) = c^*(r, m)\mathrm{vol}(S^{2m+1})$ *where*

$$c^*(m; r) = \frac{m}{\pi} \int_0^{\pi} I(r, m, s)(\sin s)^{2m-1}\, ds$$

*with*

$$\begin{aligned} I(r, m, s) \\ = 4^m \int_0^1 u^{(2m-1)/2}(1-u)^{-1/2}(1-u+r^2\cos^2 s\, u)^{(2m-1)/2}(1-u+r^2 u)^{1/2}\, du. \end{aligned}$$

*In particular for $r = 1$,*

$$\mathrm{Vol}(P) = 4^{2m}\binom{4m}{2m}^{-1} \mathrm{vol}(S^{2m+1}).$$

***Proof*** We choose for $S^{2m+1}$ the usual spherical coordinates $\{\varphi_1, \ldots, \varphi_{n-2}, s, t\}$ modulo the identification of the basis $\{e_1, \ldots, e_{n-1}, b, a\}$ with the canonical one. With this choice, $f_a = \cos t$, $f_b = \sin t \cos s$ and $f_P$ is a function that only depends on $s, t$ that we will represent by $f(r, m, s, t)$. Then by (5.14)

$$f(r, m, s, t) = \left(1 + \frac{r^2(1 - \cos t)\cos^2 s}{1 + \cos t}\right)^{(2m-1)/2} \left(1 + \frac{r^2(1 - \cos t)}{1 + \cos t}\right)^{1/2}$$

and

$$\mathrm{Vol}(P) = \mathrm{vol}(S^{2m-1}) \int_0^{\pi} \int_0^{\pi} f(r, m, s, t)(\sin s)^{2m-1}(\sin t)^{2m}\, ds\, dt.$$

If we define $I(r, m, s) := \int_0^{\pi} f(r, m, s, t)(\sin t)^{2m}\, dt$, and after the change of variable $u = \frac{1}{2}(1 - \cos t)$

$$I(r, m, s)$$
$$= 4^m \int_0^1 u^{(2m-1)/2}(1 - u)^{-1/2}(1 - u + r^2 \cos^2 s\, u)^{(2m-1)/2}(1 - u + r^2 u)^{1/2}\, du \tag{5.23}$$

and we obtain the first assertion using that $\mathrm{vol}(S^{2m+1}) = \frac{\pi}{m}\mathrm{vol}(S^{2m-1})$, as follows from (5.5).

If we assume $r = 1$, the integral $I(1, m, s)$ can be written in terms of the $\Gamma$ function and of the hypergeometric function ${}_2F_1$ as follows

$$\begin{aligned} I(1, m, s) &= 4^m \int_0^1 u^{(2m-1)/2}(1 - u)^{-1/2}((1 - u) + \cos^2 s\, u)^{(2m-1)/2}\, du \\ &= \frac{(-4)^m \pi^{3/2}}{\Gamma(1/2 - m)\Gamma(1 + m)}\, {}_2F_1(1/2 - m, 1/2 + m, 1 + m; \sin^2 s). \end{aligned} \tag{5.24}$$

On the other hand

$$\int_0^{\pi} {}_2F_1(1/2 - m, 1/2 + m, 1 + m; \sin^2 s)(\sin s)^{2m-1}\, ds = \frac{m\pi^{1/2}\Gamma(m)^2}{\Gamma(1/2 + 2m)}$$

that with (5.23) and (5.24) gives

$$\begin{aligned} c^*(m; 1) &= \left(\frac{(-4)^m m\pi^{1/2}}{\Gamma(1/2 - m)\Gamma(1 + m)}\right)\left(\frac{m\pi^{1/2}\Gamma(m)^2}{\Gamma(1/2 + 2m)}\right) \\ &= \pi^{1/2}\frac{\Gamma(1 + 2m)}{\Gamma(1/2 + 2m)} = 4^{2m}\binom{4m}{2m}^{-1}. \end{aligned}$$

□

The function $I(r, m, s)$ for general $r > 0$ can be expressed in terms of the generalised hypergeometric series ${}_3F_2$ as it has been computed in [11].

**Proposition 5.5** *Let $V$ be a unit vector field on $S^{2m+1}$. The variation of unit vector fields obtained by parallel transport of $V$ along the integral curves of $\overline{R}^a$ for $a \in \mathbf{R}^{n+1}$ with $\|a\| = 1$ verifies*

$$\lim_{t\to\infty} V_t^a = P^{-a,b} \quad \text{with} \quad b = V(-a) \quad \text{and}$$

$$\lim_{t\to-\infty} V_t^a = P^{a,b} \quad \text{with} \quad b = V(a).$$

*By construction the variational vector field $A(p) = (V_t^a(p))'(0)$ is $\nabla_{\overline{R}^a} V$.*

***Proof*** With the same notation as in Proposition 5.1, let us represent by $\tau_{t+t_p}^{t_p}$ the parallel transport along the curve $c_p$ from $t + t_p$ to $t_p$ and let us define

$$V_t^a(p) = \tau_{t+t_p}^{t_p}(V(c_p(t + t_p))).$$

Due to the vanishing of $\overline{R}^a$ at $p = \pm a$, all the vector fields of the variation verify $V_t^a(a) = V(a)$ and $V_t^a(-a) = V(-a)$.

For $p \neq \pm a$, by definition $V_t^a(p) = W(t_p)$ where $W(u)$ is the parallel vector field along the curve $c_p(u)$ with initial condition $W(u_0) = V(c_p(u_0))$ with $u_0 = t + t_p$. It's easy to see that $W$ must be the solution of the ODE

$$\frac{dW}{du}(u) = \frac{k(t,p)}{\cosh(u)}\Big( - \tanh(u)a + \frac{1}{\cosh(u)} \frac{p - f_a(p)a}{\sqrt{1 - f_a(p)^2}}\Big)$$

with $k(t, p) = \cosh(t + t_p)g(a, V(c_p(t + t_p)))$ from where we obtain

$$W(u) = k(t,p)\Big(\frac{1}{\cosh(u)}a + \tanh(u)\frac{p - f_a(p)a}{\sqrt{1 - f_a(p)^2}}\Big) + K(t,p)$$

with $K(t, p)$ given by the initial condition $W(t + t_p) = V(c_p(t + t_p))$ and then

$$\begin{aligned} V_t^a(p) &= k(t,p)\Big(\frac{1}{\cosh(t_p)}a + \tanh(t_p)\frac{p - f_a(p)a}{\sqrt{1 - f_a(p)^2}}\Big) + K(t,p) \\ &= V(c_p(t + t_p)) + k(t,p)X(t,p) \end{aligned} \tag{5.25}$$

where

$$X(t,p) = \Big(\frac{1}{\cosh(t_p)} - \frac{1}{\cosh(t + t_p)}\Big)a + (\tanh(t_p) - \tanh(t + t_p))\frac{p - f_a(p)a}{\sqrt{1 - f_a(p)^2}}.$$

As pointed out in Proposition 5.1, $\tanh(t_p) = -f_a(p)$ and then

$$\lim_{t\to-\infty} X(t,p) = \sqrt{1-f_a(p)^2}a + (1-f_a)\frac{p-f_a(p)a}{\sqrt{1-f_a(p)^2}}$$
$$= \frac{1-f_a}{\sqrt{1-f_a(p)^2}}(a+p)$$
$$\lim_{t\to\infty} X(t,p) = \sqrt{1-f_a(p)^2}a - (1+f_a)\frac{p-f_a(p)a}{\sqrt{1-f_a(p)^2}}$$
$$= \frac{1+f_a}{\sqrt{1-f_a(p)^2}}(a-p). \tag{5.26}$$

On the other hand, if we use de fact that $g(V(c_p(t+t_p)), c_p(t+t_p)) = 0$ jointly with the expression of $c_p$ obtained in Proposition 5.1

$$k(t,p)\tanh(t+t_p) = g(\sinh(t+t_p)a, V(c_p(t+t_p)))$$
$$= g\Big(\frac{p-f_a(p)a}{\sqrt{1-f_a(p)^2}}, V(c_p(t+t_p))\Big)$$

and taking the limits

$$\lim_{t\to-\infty} k(t,p) = \lim_{t\to-\infty} -k(t,p)\tanh(t+t_p) = -\frac{g(p,V(a))}{\sqrt{1-f_a(p)^2}}$$

$$\lim_{t\to\infty} k(t,p) = \lim_{t\to\infty} k(t,p)\tanh(t+t_p) = \frac{g(p,V(-a))}{\sqrt{1-f_a(p)^2}}. \tag{5.27}$$

From (5.25), (5.26), (5.27) and (5.7)

$$\lim_{t\to\infty} V_t^a(p) = V(-a) - \frac{g(p,V(-a))}{1-f_a(p)}(-a+p) = P^{-a,b}(p) \quad \text{with} \quad b = V(-a).$$

Analogously for the limit when $t \to -\infty$. □

**Corollary 5.1** *Let $H$ be a unit Hopf vector field on $S^{2m+1}$ and $H_t^a$ the variation of smooth unit vector fields constructed as in the Proposition above. For each $r < r_0(m)$, with $r_0(m) = \sqrt{2m-3}$, there is $t_0$ such that* $\mathrm{Vol}(rH_t^a) < \mathrm{Vol}(rH)$ *for all $t \le t_0$.*

***Proof*** By construction, the variation $rH_t^a$ issues from $rH$ and has variational vector field $r\nabla_{\overline{R}^a}H$. Since $\overline{R}^a = -a + g(a,N)N$, by (3.1) we know that

$$\nabla_{\overline{R}^a}H = J(-a+g(a,N)N+g(a,H)H) = -Ja + g(a,N)H - g(a,H)N$$

which is one of the instability directions at Hopf vector fields $A_b$ with $b = -Ja$ when $r < r_0(m)$ as shown in Theorem 3.1 and therefore for $t$ small enough the volume of the vector fields in the variation is smaller than the volume of $rH$. □

In [78] the author considered a different variation of a smooth unit vector field $V$ converging to a parallel translation vector field $P$, namely she described it as the family $W_t$ obtained by conformal stretching, i. e. $W_t = \phi_*^{-1}(t\phi_*(V))$ where $\phi$ is the stereographic projection. These vector fields were used to exhibit smooth unit vector fields with less volume than the Hopf vector fields for $m > 1$, after showing that $c^*(m; 1) < 2^m$.

It's not difficult to check that the variation that we have constructed in Proposition 5.5 is related with $W_t$ by a change of variable $t \to e^t$.

**Proposition 5.6** *The volume of the parallel transport vector fields of length $r$ on $S^{2m+1}$ can be approximated by the volume of smooth vector fields of the same constant length. Therefore, when $r$ satisfies $c^*(m; r) < (1 + r^2)^m$ the Hopf vector fields $rH$ can't be the minimisers of the volume functional among smooth vector fields; the inequality occurs in particular for all $m > 1$ and $r = 1$.*

***Proof*** Let's start with a smooth unit vector field $V$ on $S^{2m+1}$ and let's construct the variation $V_t^a$ converging to $P = P^{-a,b}$ with $b = V(-a)$. For each $0 < \varepsilon < \pi$ we consider the compact set $C_\varepsilon = \overline{B}_\varepsilon \setminus B_{\varepsilon/2}$, where $B_\varepsilon$, $B_{\varepsilon/2}$ are the geodesic balls of center $a$ and radius $\varepsilon$ and $\varepsilon/2$, respectively. There is $t_\varepsilon$ such that for $t \geq t_\varepsilon$ we have $\|V_t^a - P\| < \varepsilon$ on $C_\varepsilon$. Let $h : \mathbf{R} \to \mathbf{R}$ be a smooth increasing function $0 \leq h \leq 1$ which vanishes on $]-\infty, 1/2]$, is equal to 1 on $[1, \infty[$ and such that $h' < 2$, then the unit vector field

$$W_\varepsilon = \begin{cases} P \quad \text{on} \quad S^{2m+1} \setminus B_\varepsilon, \\ \left(h(\frac{\rho}{\varepsilon})P + (1 - h(\frac{\rho}{\varepsilon}))V_{t_\varepsilon}^a\right)/\|h(\frac{\rho}{\varepsilon})P + (1 - h(\frac{\rho}{\varepsilon}))V_{t_\varepsilon}^a\| \quad \text{on} \quad C_\varepsilon \quad \text{and} \\ V_{t_\varepsilon}^a \quad \text{on} \quad B_{\varepsilon/2} \end{cases}$$

where $\rho$ is the geodesic distance to $a$. The vector field $W_\varepsilon$ is smooth provided that $\|h(\frac{\rho}{\varepsilon})P + (1 - h(\frac{\rho}{\varepsilon}))V_{t_\varepsilon}^a\|$ never vanishes, for what we will assume $\varepsilon < 2$. By construction, not only the family $rW_\varepsilon$ converges pointwise to $rP$ on $S^{2m+1} \setminus \{a\}$ as $\varepsilon \to 0$ but also $f_{rW_\varepsilon}$ tends to $f_{rP}$ pointwise and then

$$\lim_{\varepsilon \to 0} \int_{S^{2m+1} \setminus B_\varepsilon} f_{rW_\varepsilon} \mathrm{dv}_g = \lim_{\varepsilon \to 0} \int_{S^{2m+1} \setminus B_\varepsilon} f_{rP} \mathrm{dv}_g = \mathrm{Vol}(rP).$$

On the other hand, on $B_\varepsilon$ the covariant derivative $\nabla W_\varepsilon$ is at most of order $\frac{1}{\varepsilon}$ and then $f_{rW_\varepsilon}$ is at most of order $\frac{1}{\varepsilon^{2m}}$ and then

$$\lim_{\varepsilon \to 0} \int_{B_\varepsilon} f_{rW_\varepsilon} \mathrm{dv}_g = 0$$

which ends the proof of the first statement. For the second one, accordingly with Proposition 5.4 we only need to show that

$$4^{2m} < \binom{4m}{2m} 2^m.$$

But for $m = 2$ , $4^3 < \binom{8}{4} = 70$ and the proof follows by induction on $m \geq 2$. □

**Proposition 5.7** *For each $m > 1$ there is a value $r_1(m)$ such that if $r > r_1(m)$ then $\mathrm{Vol}(rH) < \mathrm{Vol}(rP^{a,b})$.*

***Proof*** From Corollary 4.1, we only need to show that the twisting of the unit parallel transport vector fields verify $T(P^{a,b}) > T(H) = \mathrm{vol}(M)$. Using the expressions (5.10) and (5.11) it is easy to see that

$$\sigma_{2m}((\nabla P)^t \nabla P)) = \frac{1 - f_a^2}{(1 + f_a)^2} \Big( \frac{f_b^2}{(1 + f_a)^2} \Big)^{2m-1}$$

and if we choose coordinates on the sphere as described in Proposition 5.4 then $f_a = \cos t$, $f_b = \sin t \cos s$ and

$$T(P) = \mathrm{vol}(S^{2m-1}) \int_0^\pi \int_0^\pi \frac{(\sin t)^{2m} (|\cos s|)^{2m-1}}{(1 + \cos t)^{2m}} (\sin s)^{2m-1} (\sin t)^{2m} \, ds \, dt. \tag{5.28}$$

The double integral splits into two integrals

$$\int_0^\pi \frac{(\sin t)^{4m}}{(1 + \cos t)^{2m}} dt = \frac{4^m \sqrt{\pi} \Gamma(1/2 + 2m)}{\Gamma(1 + 2m)}$$

$$\int_0^\pi (|\cos s|)^{2m-1} (\sin s)^{2m-1} \, ds = \frac{\sqrt{\pi} \Gamma(m)}{\Gamma(1/2 + m)}$$

that plugged in (5.28) give

$$T(P) = \frac{4^m \pi \Gamma(1/2 + 2m) \Gamma(m)}{\Gamma(1/2 + m) \Gamma(1 + 2m)} \mathrm{vol}(S^{2m-1}).$$

Since $\mathrm{vol}(S^{2m-1}) = \frac{m}{\pi} \mathrm{vol}(S^{2m+1})$, the twisting of $P$ verifies

$$T(P) = \binom{2m}{m}^{-1} \binom{4m}{2m} \mathrm{vol}(S^{2m+1}) > T(H)$$

as we wanted to show. □

## 5.3 The Main Open Problem

The main open problem is to compute the infimum of the volume of smooth vector fields of constant length on odd-dimensional spheres of dimension greater or equal to 5. More precisely, for each $m > 1$ and $r > 0$, the question is to find the value

$$\mathcal{V}(m,r) = \inf\{\mathrm{Vol}(V)\ ;\ V \in \Gamma^{\infty}(T^r(S^{2m+1}))\}$$

and to characterise the smooth vector fields minimising the functional, if the infimum is attained.

We summarise the partial results that we have shown in the text that can help to understand better the problem in view of a solution. On one hand the volume of the radial vector fields $\mathrm{Vol}(R^a) = c(m;r)\mathrm{vol}(S^{2m+1})$ is a lower bound of the volume functional that can't be attained by smooth vector fields (see Theorem 4.2 and Proposition 5.2) but it's not known if this value can be approached by the volume of smooth vector fields.

On the other hand, the volume of the Hopf vector fields of length $r$ is an upper bound of $\mathcal{V}(m,r)$ and since it's equal to $(1+r^2)^m\mathrm{vol}(S^{2m+1})$ by Proposition 3.2, then

$$c(m;r)\mathrm{vol}(S^{2m+1}) \leq \mathcal{V}(m,r) \leq (1+r^2)^m\mathrm{vol}(S^{2m+1}).$$

To have an idea of the gap between the two bounds, we only need to remember that

$$c(m;r) = \sum_{k=0}^{m} \frac{\binom{m}{k}^2}{\binom{2m}{2k}} r^{2k}$$

and then, although $c(1;r) = 1 + r^2$, the gap increases with $m$ quickly. For instance $c(2;r) = 1 + \frac{2}{3}r^2 + r^4$.

A different estimate of $\mathcal{V}(m,r)$ is related with the parallel transport vector fields $P^{a,b}$ since, although they are not smoothly defined in the whole sphere, their volume can be approximated by the volume of smooth vector fields of constant length, as it has been seen in Proposition 5.6 and consequently

$$\mathcal{V}(m;r) \leq c^*(m;r)\mathrm{vol}(S^{2m+1}).$$

In the particular case of $r = 1$, this fact was used in [78] to exhibit smooth unit vector fields with less volume than the Hopf vector fields for $m > 1$, after showing that $c^*(m;1) < 2^m$ (see Proposition 5.6). The result led the author to conjecture that $\mathcal{V}(m;1) = c^*(m;1)\mathrm{vol}(S^{2m+1})$. This conjecture is still open but D. L. Johnson and P. Smith in [68] have shown that the unit vector fields $P^{a,b}$ are not volume minimisers if one considers a set of sections larger than $\Gamma^{\infty}(T^1(S^{2m+1}))$, namely the set of rectifiable sections.

In view of Proposition 5.3 we can assert that, for $r \neq 1$, the lack of minimality of the parallel transport vector fields implies that in fact $c^*(m;r)\text{vol}(S^{2m+1})$ can't be the infimum. Still this value can be useful to obtain an upper bound of $\mathcal{V}(m;r)$.

Whether the upper bound of $\mathcal{V}(m;r)$ coming from parallel transport vector fields is better than the one coming from Hopf vector fields depends on the length $r$. Indeed, as we have shown in Proposition 5.7, for each $m > 1$ there is a value $r_1(m)$ such that if $r > r_1(m)$ then $(1+r^2)^m < c^*(m;r)$ and consequently if the length $r$ is big enough the Hopf vector fields have less volume than the parallel transport ones. We don't know the explicit values of $r_1(m)$ but, as noted in [11], numerical computations using Proposition 5.4 give for example the approximate values

$$r_1(2) = \sqrt{1,815},\ r_1(3) = \sqrt{5,563},\ r_1(10) = \sqrt{26,29},\ r_1(100) = \sqrt{286,1}$$

that suggest that $r_0(m) = \sqrt{2m-3} < r_1(m)$.

Using the instability results, in particular Theorem 3.1, we know that the Hopf vector fields can't be minimisers for $r < r_0(m)$ and in particular we can consider the variations of Hopf vector fields as in Corollary 5.1 and to compute their volumes to obtain better upper bounds. These variations starting at Hopf vector fields are decreasing volume for small $t$, by construction, and converge to a parallel transport vector field that has less volume than the Hopf vector fields but we don't know if they are volume decreasing for all $t$. With the exception of $r = 1$, the lack of minimality of the parallel transport vector fields implies that $\mathcal{V}(m;r) < c^*(m;r)\text{vol}(S^{2m+1})$.

In conclusion, for $r \geq r_1(m)$ the Hopf vector fields are stable critical points and by now there are no examples of smooth vector fields of length $r$ with less volume.

In [32], D. DeTurck, H. Gluck and P.A. Storm have shown that the unit Hopf vector field is up to isometries of domain and range the unique Lipschitz constant minimiser in its homotopy class. Here the Lipschitz constant of a continuous map $f : X \to Y$ between metric spaces is the smallest number $c \geq 0$ such that $d(f(x), f(y)) \leq cd(x,y)$ for all points $x, y \in X$. This minimising property is related with the problem about the volume since, as the authors point out, if a smooth submanifold of the Riemannian manifold is a volume minimising cycle in its homology class, then the inclusion map is a Lipschitz constant minimiser in its homotopy class.

Also related with the main open problem is the work by Brito et al. [23] where it is shown that if $M \subset S^{2m+1}$ is a domain with boundary $\partial M \neq \emptyset$ then the restriction to $M$ of the Hopf vector fields minimise the volume among those unit vector fields defined on $M$ that coincide with the Hopf vector field on $\partial M$.

If we consider the same open problem of finding the infimum of the volume of smooth vector fields of constant length for a complete $(2m+1)$-dimensional manifold $M$ of constant curvature 1, different from the sphere, the situation is more satisfactory because on one hand we know that Hopf vector fields are stable for all $r$ (Theorem 3.4) and on the other hand, by Corollaries 4.1 and 4.2, for any unit vector field $V$ which is not a Hopf vector field there is $r_V > 0$ such that $\text{Vol}(rV) >$

$\mathrm{Vol}(rH)$ for all $r < r_V$ and if its twisting verifies $T(V) \neq T(H)$ there is $\overline{r}_V > 0$ such that $\mathrm{Vol}(rV) > \mathrm{Vol}(rH)$ for all $r > \overline{r}_V$. Moreover there are no examples of smooth vector fields of length $r$ with less volume than the Hopf vector fields. In view of these results and their proofs we are led to the following

*Conjecture 5.1* On a complete $(2m + 1)$-dimensional manifold $M$ of constant curvature, different from the sphere, the infimum of the volume of vector fields of constant length is attained exactly by Hopf vector fields.

## 5.4 Area Minimising Vector Fields on the 2-Sphere

For a manifold $M$ without smooth vector fields of constant length, we can look for those that minimise the volume among the most regular ones. We can take different points of view to define the meaning of *most regular*. One possibility is to look at vector fields smooth in the complementary of a single point, although we found more interesting to consider the class of constant length vector fields *without boundary*, due to their relation with the minimal submanifolds of the sphere tangent bundles of $M$. The notion was defined in our joint work with Borrelli [12] as follows:

**Definition 5.3** Let $U$ be a dense open subset of a closed manifold $M$, let $V : U \to T^rM$ be a smooth vector field and let's represent by $\Sigma_V$ the closure of $V(U) \subset T^rM$. The vector field $V$ is said to be without boundary if $\Sigma_V$ is an embedded submanifold without boundary.

In [12] we have shown that for the 2-dimensional sphere the parallel transport vector fields are the minimisers both among vector fields without boundary and among those defined in the complementary of a single point.

The key point of these results is that the geometry of the unit tangent bundle of $S^2$ with the Sasaki metric is well known, as W. Klingenberg and T. Sasaki proved in [70] that it is isometric to the real projective space $\mathbf{R}P^3$ obtained as the quotient of the sphere $S^3(2)$ of radius 2. In [12] we have shown that the sphere tangent bundles $T^rS^2$ for $r > 0$ are homothetic to the projective space with a metric $\overline{g}_\mu$ obtained as a quotient of a Berger sphere.

**Proposition 5.8** *The manifold $(T^rS^2, g^S)$ is isometric to the projective space obtained by the quotient of $(S^3, 4g_{r^2})$.*

***Proof*** For an element $(x, v) \in T^1S^2$ we can consider the isometry of $\mathbf{R}^3$ denoted by $\psi(x, v)$ and defined as

$$\psi(x, v)(e_1) = x, \qquad \psi(x, v)(e_2) = v, \qquad \psi(x, v)(e_3) = x \wedge v,$$

where $\{e_1, e_2, e_3\}$ is the usual base. The map

$$\psi : (T^1S^2, g^S) \to (SO(3), \frac{1}{2} <\ ,\ >)$$

defined in this way is an isometry, when $<\ ,\ >$ is the usual bi-invariant metric on $SO(3)$ given by $< A, B >= \mathrm{tr}(A^t B)$ for all $A, B \in \mathfrak{so}(3)$. Let's prove it.

Let $(x, v) \in T^1S^2$ that we can assume without lost of generality to be $(e_1, e_2)$ and let $\xi \in T_{(x,v)}(T^1S^2)$. If $V(t) = (x(t), v(t))$ is a curve in $T^1S^2$ with $x(0) = e_1$, $v(0) = e_2$ and $V'(0) = \xi$, then $\psi_{*|(e_1,e_2)}(\xi) \in \mathfrak{so}(3)$ is given by

$$\psi_{*|(e_1,e_2)}(\xi)(e_1) = x'(0),$$
$$\psi_{*|(e_1,e_2)}(\xi)(e_2) = v'(0),$$
$$\psi_{*|(e_1,e_2)}(\xi)(e_3) = x'(0) \wedge e_2 + e_1 \wedge v'(0).$$

Since $(x, v) \in T^1S^2$ we have that $x_1'(0) = 0$, $v_2'(0) = 0$ and $x_2'(0) + v_1'(0) = 0$ from where

$$\frac{1}{2} < \psi_{*|(e_1,e_2)}(\xi), \psi_{*|(e_1,e_2)}(\xi) >= |x_2'(0)|^2 + |x_3'(0)|^2 + |v_3'(0)|^2.$$

On the other hand,

$$g^S(\xi, \xi) = g(x'(0), x'(0)) + g(\frac{\nabla v}{dt}(0), \frac{\nabla v}{dt}(0)) = |x_2'(0)|^2 + |x_3'(0)|^2 + |v_3'(0)|^2,$$

where for the second equality we have used that

$$\frac{\nabla v}{dt}(0) = v'(0) - g(v'(0), x(0))x(0) = v_3'(0)e_3.$$

The next step is to prove that, if we denote by $\overline{g}$ the quotient metric and by $\overline{\Phi}$ the pass to the quotient of the Euler parametric representation of $SO(3)$, then $\overline{\Phi} : (\mathbf{R}P^3, 8\overline{g}) \to (SO(3), <\ ,\ >)$ is an isometry.

For the proof it's convenient to identify $\mathbf{R}^4$ with the quaternions $\mathbf{H}$, $\mathbf{R}^3$ with the imaginary quaternions $Im\ \mathbf{H}$ and then $SO(3) = SO(Im\ \mathbf{H})$. The Euler parametric representation $\Phi : S^3 \subset \mathbf{H} \to SO(3)$ can be expressed as the group epimorphism from the subgroup $S^3$ of $\mathbf{H}$ onto $SO(Im\ \mathbf{H})$ such that for $q \in S^3$ its image $\Phi(q)$ is given by $\Phi(q)(u) = q^{-1}uq$. By passing to the quotient, $\overline{\Phi}$ is a diffeomorphism from the projective space onto $SO(3)$.

The differential of $\Phi$ at $e = 1 \in S^3$ is then a linear map from $T_eS^3 = Im\ \mathbf{H}$ to the Lie algebra $\mathfrak{so}(Im\ \mathbf{H})$ that is given by $\Phi_{*|e}(X)(u) = Xu - uX$ for all $X \in T_eS^3$ and $u \in Im\ \mathbf{H}$. By straightforward computation one obtains

$$< \Phi_{*|e}(X), \Phi_{*|e}(X) >= 8g(X, X).$$

The composition $\Psi = \psi^{-1} \circ \overline{\Phi}$ is then an isometry between $(\mathbf{R}P^3, 4\overline{g})$ and $(T^1S^2, g^S)$. Moreover by definition $\pi(\Psi(e^{i\theta}q)) = \Phi(e^{i\theta}q)(i) = \Phi(q)(i) = \pi(\Psi(q))$ and therefore $\Psi$ preserves the natural structures of $S^1$-bundle over $S^2$ of both spaces $S^3$ and $T^1S^2$.

To finish the proof we only need to see how $(T^rS^2, g^S)$ is related with $(T^1S^2, g^S)$. Let $h$ be the map from $T^1S^2$ to $T^rS^2$ given by $h(x, v) = (x, rv)$ then $h_{*|(x,v)}(V'(0)) = (x'(0), rv'(0))$ and we have

$$g^S(h_{*|(x,v)}(V'(0)), h_{*|(x,v)}(V'(0))) = g(x'(0), x'(0)) + r^2 g(\frac{\nabla v}{dt}(0), \frac{\nabla v}{dt}(0)).$$

So, $(T^rS^2, g^S)$ is isometric to $T^1S^2$ endowed with the metric obtained by deforming the Sasaki metric in the vertical direction by a factor $r^2$. Since $\Psi$ preserves the bundle structures, the map $h^{-1} \circ \Psi$ is an isometry from the projective space considered with the metric $4\overline{g}_\mu$, where $g_\mu$ is the Berger metric with $\mu = r^2$, onto $(T^rS^2, g^S)$. □

**Proposition 5.9** *The parallel transport vector fields of length $r$ on $S^2$ are without boundary, in particular the submanifold $\Sigma_P$ is mapped by the isometry $\Psi$ onto an $\mathbf{R}P^2$ obtained as the quotient of an equatorial sphere of $S^3$ and conversely every such a projective plane in $\mathbf{R}P^3$ comes from a parallel transport vector field.*

***Proof*** By the construction of $\Psi$, it suffices to show the result for $r = 1$ and, modulo a change of the orthonormal base of $\mathbf{R}^3$ if necessary, we can assume that $P = P^{a,b}$ with $a = (0, 0, 1)$ and $b = (0, 1, 0)$. In view of (5.7)

$$P(x, y, z) = \Big(-\frac{xy}{1+z}, 1 - \frac{y^2}{1+z}, -y\Big). \tag{5.29}$$

We are going to see that $\psi(\Sigma_P) = \Phi(\{(\kappa, \lambda, \mu, -\lambda) \in S^3\})$. For the proof it is convenient to write the Euler parametric representation $\Phi$ in coordinates; if $q = (\kappa, \lambda, \mu, \nu) \in S^3$ its image $\Phi(q) \in SO(3)$ is the isometry of matrix

$$\begin{pmatrix} \kappa^2 + \lambda^2 - \mu^2 - \nu^2 & 2\kappa\nu + 2\lambda\mu & -2\kappa\mu + 2\lambda\nu \\ -2\kappa\nu + 2\lambda\mu & \kappa^2 - \lambda^2 + \mu^2 - \nu^2 & 2\kappa\lambda + 2\mu\nu \\ 2\kappa\mu + 2\lambda\nu & -2\kappa\lambda + 2\mu\nu & \kappa^2 - \lambda^2 - \mu^2 + \nu^2 \end{pmatrix}.$$

Then $\Phi(q) \in \psi(\Sigma_P \setminus \partial\Sigma_P)$ if and only if there exists $(x, y, z) \in S^2$ with $z \neq -1$ such that

$$(\kappa^2 + \lambda^2 - \mu^2 - \nu^2, -2\kappa\nu + 2\lambda\mu, 2\kappa\mu + 2\lambda\nu) = (x, y, z) \tag{5.30}$$

and

$$(2\kappa\nu + 2\lambda\mu, \kappa^2 - \lambda^2 + \mu^2 - \nu^2, -2\kappa\lambda + 2\mu\nu) = \left(-\frac{xy}{1+z}, 1 - \frac{y^2}{1+z}, -y\right). \tag{5.31}$$

If $(\kappa, \lambda, \mu, -\lambda) \in S^3$, equality (5.30) implies that $x = \kappa^2 - \mu^2$, $y = 2\lambda(\kappa + \mu)$ and $z + 1 = (\kappa + \mu)^2$ and then it's immediate that $q$ verifies (5.31). Now, let us assume that $q = (\kappa, \lambda, \mu, \nu) \in S^3$ satisfies both equations but $\lambda \neq -\nu$. Then, the relation between the second coordinate of equality (5.30) and the third coordinate of equality (5.31) gives

$$(\mu - \kappa)(\lambda + \nu) = 0 \tag{5.32}$$

which would imply $\mu = \kappa$ and then Eqs. (5.30) and (5.31) would become

$$(\lambda^2 - \nu^2, 2\kappa(-\nu + \lambda), 2\kappa^2 + 2\lambda\nu) = (x, y, z) \tag{5.33}$$

and

$$(2\kappa(\nu + \lambda), 2\kappa^2 - \lambda^2 - \nu^2, 2\kappa(-\lambda + \nu)) = \left(-\frac{xy}{1+z}, 1 - \frac{y^2}{1+z}, -y\right), \tag{5.34}$$

respectively. Then

$$2\kappa(\nu + \lambda) = -\frac{xy}{1+z} = -\frac{(\lambda^2 - \nu^2)2\kappa(\lambda - \nu)}{1+z} \tag{5.35}$$

$$-2(\lambda^2 + \nu^2) = -\frac{y^2}{1+z} = -\frac{(\lambda^2 - \nu^2)2\kappa(\lambda - \nu)}{1+z}. \tag{5.36}$$

By (5.36) $\kappa \neq 0$ and as we are assuming that $\lambda + \nu \neq 0$ then (5.35) would imply that $-(\lambda - \nu)^2 = 1 + z$ which is impossible, since $1 + z > 0$, which ends the proof. □

*Remark* For the particular case of $\mathbf{R}P^3$ with the round metric these equatorial projective planes are totally geodesic submanifolds, but in general for the projective space with a Berger metric Proposition above, jointly with Proposition 5.3, show that they are minimal submanifolds. In [12] we provided a direct proof of this result.

The topology of the surfaces $\Sigma_V$ corresponding to vector fields without boundary of $S^2$ could be quite diverse. For instance, as it can be seen in [12], if $V$ has a finite number of singularities $k$ then $\Sigma_V$ is homeomorphic to the connected sum of a projective plane and a torus with $(k - 1)/2$ holes. Anyhow, all the surfaces $\Sigma_V$ are in the same homology class.

**Lemma 5.3** *For every unit vector field $V$ on $S^2$ without boundary the homology class with $\mathbf{Z}_2$-coefficients $[\Sigma_V] \in H_2(T^1S^2, \mathbf{Z}_2) = \mathbf{Z}_2$ is the non-trivial class.*

***Proof*** In view of Proposition 5.8, the homology group $H_2(T^1S^2, \mathbf{Z}_2)$ with $\mathbf{Z}_2$-coefficients is equal to $H_2(\mathbf{R}P^3, \mathbf{Z}_2) = \mathbf{Z}_2$ (the generator is the class of $\mathbf{R}P^2 \subset \mathbf{R}P^3$). Since $V$ is a vector field defined on an open and dense subset of $S^2$, the map $\pi_*$ induced by the projection $\pi : T^1S^2 \to S^2$ on the corresponding homology groups verifies that $\pi_*([\Sigma_V])$ is the homology class of $S^2$ and therefore $[\Sigma_V]$ cannot be the trivial class. □

**Theorem 5.1**

*(1) Unit parallel transport vector fields are the only minimisers of the area, among the unit vector fields without boundary of $S^2$.*
*(2) For $r \neq 1$, parallel transport vector fields are the only minimisers of the area, among the vector fields $V$ of length $r$ on $S^2$, without boundary and such that $\Sigma_V$ is homeomorphic to a projective plane.*

***Proof*** For the projective space with the round metric, we know that embedded totally geodesic projective planes are the only area minimising surfaces in their homology class, as it was shown by M. Berger in [1] (see also Fomenko [35]). Therefore, the first part of the theorem follows from the Lemma above, the isometry $\Psi$ of the unit tangent bundle onto the projective space exhibited in Propositions 5.8 and the fact that every equatorial $\mathbf{R}P^2$ is of the form $\Psi(\Sigma_P)$ for some parallel transport vector field $P$, proved in Proposition 5.9.

For general $r \neq 1$, all the ingredients of the previous reasoning are valid with the exception of the Berger's result about the minimisers of the area. We have to use instead our result in [42] where we have shown that the only projective planes minimally embedded in $(\mathbf{R}P^3, g_\mu)$ are the equatorial projective planes and that, moreover, the equatorial $\mathbf{R}P^2$ in $(\mathbf{R}P^3, g_\mu)$ are exactly the area minimising embedded submanifolds among those that are homeomorphic to projective planes. □

*Remark* Part 1 of the Theorem 5.1 is the only result of the book where we need to use that all the vector fields considered determine submanifolds of the sphere bundle that are in the same homology class. The reason is because in the Berger result we use in the proof this property is required

**Open Question** It would be interesting to know, when $r \neq 1$, whether the additional assumption of $\Sigma_V$ to be homeomorphic to a projective plane can be avoided. In [12] we obtained the theorem by showing that the only minimal surfaces of $T^rS^2$ homeomorphic to the projective plane arising from vector fields without boundary are of the form $\Sigma_P$. Via the isometry $\Psi$ this result is a weaker version of the above mentioned result of [42], valid for surfaces of $(\mathbf{R}P^3, g_\mu)$ that are sections of the Hopf fibration on an open and dense subset of $S^2$. The proofs on [12] and [42] follow a different method, but in both cases the restriction to surfaces homeomorphic to the projective plane seems to be essential.

As a consequence of the Theorem above, we can compute the value of the infimum of the area of vector fields on $S^2$, among those verifying its hypotheses, in terms of $r$. With the same arguments as in Proposition 5.4, the volume of a parallel transport vector field $P$ of length $r$ is

$$\mathrm{Vol}(P) = 2\pi \int_0^{\pi} \sqrt{1 + \frac{r^2(1-\cos t)}{1+\cos t}} \sin t \, dt.$$

By straightforward computation, $\mathrm{Vol}(P) = 4\pi A(r^2)$ where

$$A(\mu) = \begin{cases} 1 + \frac{\mu}{\sqrt{1-\mu}} \log\left(\frac{\sqrt{1-\mu}+1}{\sqrt{\mu}}\right) & \text{if } \mu < 1, \\ 1 + \frac{\mu}{\sqrt{\mu-1}} \arcsin\left(\frac{\sqrt{\mu-1}}{\sqrt{\mu}}\right) & \text{if } \mu > 1 \text{ and} \\ 2 \text{ if } \mu = 1. \end{cases}$$

Moreover, we have seen in Proposition 5.3 that, for all $r > 0$, every parallel transport vector field $P$ of length $r$, defined on the punctured sphere $S^2 \setminus \{-a\}$ is $r$-minimal. By Theorem 2.4, we know that the vector field orthogonal to $P$ is also $r$-minimal, but this does not provide additional information because $P^{\perp}$ is again a parallel transport vector field. The more relevant part of Theorem 2.4, in this case, is that all these vector fields are stable. Since they are vector fields defined on an open manifold, this means that they are local minimisers of the area for variations with compact support.

In Proposition 5.1, we have shown that the radial vector fields $R^a$ are $r$-minimal for all $r$ and, in view of Theorem 2.4, the vector fields of length $r$ ortogonal to them are also $r$-minimal for all $r$. These vector fields, whose trajectories are circles orthogonal to the radial geodesics, are of the form

$$C^a = \frac{r}{\sqrt{1-f_a^2}} N \wedge a.$$

With the same arguments as in Proposition 5.2, the volume of a radial vector field of length $r$ on $S^2$ and the volume of the vector field of length $r$ orthogonal to it are

$$\begin{aligned} \mathrm{Vol}(R) = \mathrm{Vol}(C) &= 2\pi \int_0^{\pi} \sqrt{1 + \frac{r^2 \cos^2 t}{1-\cos^2 t}} \sin t \, dt \\ &= 2\pi \int_0^{\pi} \sqrt{\sin^2 t + r^2 \cos^2 t} \, dt. \end{aligned}$$

In particular for $r = 1$, we have $\mathrm{Vol}(R) = \mathrm{Vol}(C) = 2\pi^2$ and $\mathrm{Vol}(P) = 8\pi$, then $\mathrm{Vol}(R) < \mathrm{Vol}(P)$ and it's easy to see, using for instance Wolfram Mathematica®, that the same inequality is true for all $r$.

This fact in not in contradiction with Theorem 5.1, because $\Sigma_R$ and $\Sigma_C$ are submanifolds of $T^r S^2$ diffeomorphic to a cylinder with boundary consisting in the two fibers over the antipodal singular points of $R$ and $C$. The two surfaces $\Sigma_R$ and

$\Sigma_C$ are mapped by $\Psi$ onto a cylinder on the projective space having as boundary the union of two fibers of the Hopf fibration. Thus, they are not vector fields without boundary.

Both, the radial vector field and their orthogonal have two singular points of index 1. Then they are neither among the most regular vector fields, of constant length on $S^2$, from the point of view of the number of singularities.

For the next result we are going to follow the same proof as in [12] for which we need the following Lemma that is obtained by straightforward computations similar to those in Theorem 2.4.

**Lemma 5.4** *If $\{E_1, E_2\}$ is an orthonormal frame on an open subset $U$ of a Riemannian manifold $(M, g)$ of dimension 2 and $V = r\cos\theta E_1 + r\sin\theta E_2$ is a unit vector field of length $r$ on $U$ then*

$$(\nabla V)^t \circ (\nabla V) = r^2 \begin{pmatrix} \beta_1^2 & \beta_1\beta_2 \\ \beta_1\beta_2 & \beta_2^2 \end{pmatrix}$$

*where $\beta_i = E_i(\theta) + g(\nabla_{E_i} E_1, E_2)$.*

**Theorem 5.2** *Among vector fields of constant length defined on $S^2$ minus one point those of least area are exactly the parallel transport vector fields.*

***Proof*** Let $V$ be a vector field with $\|V\| = r$ defined on $S^2$ minus one point, which we can assume without loss of generality to be the south pole $-a = (0, 0, -1)$, then the parallel transport vector fields $\{E_1 = P^{\,a,b}, E_2 = P^{\,a,c}\}$ with $b = (1, 0, 0)$ and $c = (0, 1, 0)$ provide an orthonormal frame on $U = S^2 \setminus \{-a\}$. The expression of $E_2$ is given by (5.29) and analogously

$$E_1(x, y, z) = \left(1 - \frac{x^2}{1+z}, -\frac{xy}{1+z}, -x\right). \tag{5.37}$$

There is a function $\theta : U \to \mathbf{R}$ such that $V = r\cos\theta E_1 + r\sin\theta E_2$ and it's easy to see that

$$\beta_1 = E_1(\theta) + \frac{y}{1+z}, \qquad \beta_2 = E_2(\theta) - \frac{x}{1+z}.$$

In view of the previous Lemma

$$\begin{aligned} f_V^2 &= 1 + r^2(\beta_1^2 + \beta_2^2) = 1 + r^2\Big(E_1(\theta) + \frac{y}{1+z}\Big)^2 + r^2\Big(E_2(\theta) - \frac{x}{1+z}\Big)^2 \\ &= 1 + r^2\Big(\frac{1-z}{1+z} + \frac{2}{1+z}(yE_1(\theta) - xE_2(\theta)) + E_1(\theta)^2 + E_2(\theta)^2\Big) \\ &= 1 + r^2\Big(\frac{1-z}{1+z} + \frac{2\sqrt{1-z^2}}{1+z}C^a(\theta) + R^a(\theta)^2 + C^a(\theta)^2\Big) \end{aligned} \tag{5.38}$$

where $R^a$, $C^a$ are the radial vector field and its orthogonal defined on $S^2 \setminus \{-a, a\}$. Using Proposition 5.4, if we represent by $P$ any of the parallel transport vector fields of length $r$ issuing from $a$, we have that

$$f_P^2 = 1 + r^2 \frac{1-z}{1+z}$$

and then, from (5.38) we can write

$$\begin{aligned} f_V^2 &= f_P^2 \Big( 1 + h_1(z) C^a(\theta) + h_2(z) R^a(\theta)^2 + h_2(z) C^a(\theta)^2 \Big) \\ &= f_P^2 \Big( \Big( 1 + \frac{h_1(z)}{2} C^a(\theta) \Big)^2 + h_2(z) R^a(\theta)^2 \\ &\quad + \Big( h_2(z) - \frac{h_1(z)^2}{4} \Big) C^a(\theta)^2 \Big) \end{aligned} \tag{5.39}$$

where

$$h_1(z) = \frac{2r^2\sqrt{1-z^2}}{(1+z) + r^2(1-z)} \quad \text{and} \quad h_2(z) = \frac{r^2(1+z)}{(1+z) + r^2(1-z)}. \tag{5.40}$$

On $S^2 \setminus \{-a, a\}$ both $h_2(z) > 0$ and $h_2(z) - \frac{h_1(z)^2}{4} > 0$, then from (5.39)

$$\begin{aligned} \mathrm{Vol}(V) &\geq \int_{S^2} f_P \left| 1 + \frac{h_1(z)}{2} C^a(\theta) \right| \mathrm{dv}_g \\ &\geq \left| \int_{S^2} f_P \left( 1 + \frac{h_1(z)}{2} C^a(\theta) \right) \mathrm{dv}_g \right|, \end{aligned}$$

with equality if and only if $\theta$ is constant, or equivalently if and only if $V$ is a parallel transport vector field.

If we take spherical coordinates $\{\varphi, t\}$ then $z = \cos t$, the functions $f_P$ and $h_1$ only depend on $t$, the vector field $C^a$ is $\frac{\partial}{\partial \varphi}$ and so the integral curves of $C^a$ are the parametric curves $t = t_0$. Therefore

$$\int_{S^2} f_P(z) h_1(z)\, C^a(\theta) \mathrm{dv}_g = \int_0^\pi f_P(\cos t) h_1(\cos t) \sin t \int_{-\pi}^{\pi} \frac{\partial \theta}{\partial \varphi} d\varphi dt$$

and since the vector field $V$ has a single singular point at $-a$ its index has to be 2, the same as the index of $P$ at $-a$. The number of turns of $V$ in the frame $\{P, P^\perp\}$

is zero and then

$$\int_{-\pi}^{\pi} \frac{\partial \theta}{\partial \varphi} d\varphi = 0,$$

from where we obtain the result. □

**Open Problem** It would be interesting to know whether a similar result is also true for other even dimensional spheres, although in this case one should restrict to unit vector fields due to Proposition 5.3.

## 5.5 Notes

### 5.5.1 Radial Vector Fields on Riemannian Manifolds

On a Riemannian manifold $(M, g)$ the vector fields that are tangent to the geodesics issuing from any given point $p$ are called radial vector fields. Equivalently, a radial vector field can be defined as the unit normal vector field of the geodesic spheres with centre at $p$ or as the gradient of the map $d(p, \cdot)$ given by the distance to $p$, up to normalisation. The maximal domain of definition is of the form $U \setminus \{p\}$ where $U$ is the open neighbourghood of $p$ where the exponential map is a diffeomorphism. More generally, given a submanifold $S \subset M$, one can also define radial vector fields in $U \setminus S$ as those that are tangent to the geodesics issuing from $S$ in a direction normal to $S$.

In [6], E. Boeckx and L. Vanhecke have shown that these vector fields are minimal in different types of manifolds, for instance when $M$ is a two-point homogeneous space. The simply connected two-point homogeneous space are: $S^n$, $\mathbf{C}P^n$, $\mathbf{H}P^n$ and $CaP^2$.

On the other hand, they have shown that if a 2-dimensional manifold has the property that every radial unit vector field is minimal, then it has constant curvature.

Also in [6], the reader can find the following result about the radial vector fields orthogonal to submanifolds.

**Proposition 5.10** *Let $(M, g)$ be a Riemannian manifold and let $S$ be a totally geodesic submanifold. The radial unit vector field issuing from $S$ is minimal in the following cases:*

*(a) $(M, g)$ has constant curvature.*
*(b) $(M, g)$ is one of the two-point homogeneous spaces with complex structure $J$ and $S$ is $J$-invariant.*
*(c) $(M, g)$ is a $2m$-dimensional Kähler manifold of constant holomorphic sectional curvature and $S$ is a $m$-dimensional anti-invariant submanifold.*

Let us recall that a Sasakian manifold has constant $\phi$-sectional curvature $c$ if and only if every plane admitting an orthonormal base of the form $\{X, \phi X\}$ has curvature equal to $c$. For this kind of manifolds, E. Boeckx and L. Vanhecke have shown in [7] the minimality of the radial unit vector field issuing from any characteristic line; i. e. those defined as the image of any integral curve of the characteristic vector field.

The same authors considered in [8] vector fields defined outside the set of critical points of isoparametric functions as the gradient divided by its norm. A real valued function $f$ on a Riemannian manifold is said isoparametric if

$$\|df\|^2 = b(f) \quad \text{and} \quad \Delta f = a(f),$$

where $a$ is a continuous function and $b$ is smooth.They have shown that for spaces of constant curvature they are minimal. With this approach, they also obtain examples of minimal unit vector fields defined by the unit vector fields normal to foliations that appear in certain homogeneous hypersurfaces of complex and quaternionic space forms.

A connected closed submanifold $S$ of a complete Riemannian manifold $M$ is said to be reflective if the geodesic reflection of $M$ in $S$ is a well-defined global isometry. If $M$ is a symmetric space, there exists another reflective submanifold associated to $S$, denoted $S^{\perp}$. Both are totally geodesic and then symmetric spaces. For this kind of submanifolds, J. Berndt, L. Vanhecke and L. Verhóczki have shown in [3] the following

**Proposition 5.11** *Let $M$ be a Riemannian symmetric space of compact or of non-compact type and let $S$ be a reflective submanifold of $M$ such that its codimension is greater than one and the rank of the complementary reflective submanifold $S^{\perp}$ is equal to one. Then the radial unit vector field tangent to the geodesics emanating perpendicularly from $S$ is minimal.*

Under the hypotheses, $S$ is the singular orbit of a cohomogeneity one action on $M$, the radial vector field is then defined in the open and dense set formed by the union of principal orbits. The authors provide a list of pairs $(M, S)$ with the required conditions, thus finding new explicit examples of minimal unit vector fields with singularities.

### 5.5.2 Minimisers of the Volume Among Unit Vector Fields with Singular Points

We have seen in Proposition 5.2 that for the unit radial vector fields on odd-dimensional spheres

$$\int_{S^{2m+1}} \|\nabla R\|^2 \mathrm{dv}_g = \left(1 + \frac{1}{2m-1}\right) \mathrm{vol}(S^{2m+1})$$

and then the value coincides with the lower bound in Proposition 4.4. Although the proof we have given of this proposition is valid only for smooth vector fields, it's not difficult to see as it is done in [19] that to obtain

$$\int_{S^{2m+1}} \|\nabla R\|^2 \mathrm{dv}_g \leq \int_{S^{2m+1}} \|\nabla V\|^2 \mathrm{dv}_g$$

it suffices for the unit vector field $V$ to be smooth outside a finite number of singular points. Consequently the radial unit vector fields minimise the energy among this larger family of vector fields.

In what concerns the volume functional, as far as we know the best result has been obtained by F. G. B. Brito, A. O. Gomes and I. Gonçalves in [24] where they prove

**Proposition 5.12** *Let $V$ be a unit vector field on $S^{2m+1}$ with exactly two antipodal singular points and let's represent by $i_V$ the absolute value of the index of $V$ at any of them, then* $\mathrm{Vol}(V) \geq i_V \mathrm{Vol}(R)$.

The particular case of $S^3$ was shown by F. G. B. Brito, P. M. Chacón and D. L. Johnson in [22] where the authors also treated the case of $S^2$ getting.

**Proposition 5.13** *Let $V$ be a unit smooth vector field on $S^2 \setminus \{a, -a\}$ and let's represent by $|i_V(a)|$ the absolute value of the index of $V$ at $a$ then*

$$\mathrm{Vol}(V) \geq \frac{1}{2}(|i_V(a)| + |i_V(-a)| + \pi - 2)\mathrm{vol}(S^2). \tag{5.41}$$

In particular for $V = R^a$ and $V = C^a$, where $C^a$ is the unit vector field tangent to the parallels, $|i_V(a)| + |i_V(-a)| = 2$ and $\mathrm{Vol}(R^a) = \mathrm{Vol}(C^a) = \frac{\pi}{2}\mathrm{vol}(S^2)$. Since $i_V(a) + i_V(-a) = 2$ for all unit $V$ smooth on $S^2 \setminus \{a, -a\}$ we can conclude that

**Proposition 5.14** *Unit radial vector fields and their orthogonal complementary are those of least area among unit vector fields on $S^2$ with two antipodal singularities.*

In [26], F. G. B. Brito, J. Conrado, I. Gonçalves and A. V. Nicoli have considered for each $k \in \mathbf{N}$, with $k > 1$, the class of unit smooth vector fields with exactly two antipodal singularities of index $k$ and $2 - k$ and they have found that the minimum of the volume restricted to each class is $\pi L(k)$ where $L(k)$ is the length of the ellipse $\frac{x^2}{k^2} + \frac{y^2}{(2-k)^2} = 1$. Moreover, for each $k$ they have computed, and described geometrically, the vector fields which are area minimising in the corresponding class.

The same authors jointly with G. Nunes have shown, in the recent paper [27], that if $k$ is even then the closure of the image of any area minimising vector field is a minimally immersed Klein bottle in $T^1S^2$.

From a different point of view, D. L. Johnson and P. Smith in a series of papers [66–68] have considered the volume functional extended to unit vector fields with weaker condition than smoothness, namely vector fields that are rectifiable currents, so that the compacity conditions of the enlarged space ensure the existence of minimisers. More generally they have studied also the volume of rectifiable sections.

In particular, they have proved in [67] that for each homology class of rectifiable sections of the unit tangent bundle there is a volume minimising rectifiable section of class $C^1$ over an open dense subset of $M$. In [68] they have shown some properties of these minimisers that exclude unit parallel transport vector fields for being the minimisers of the enlarged problem.

With the same technics, P. M. Chacón and D. L. Johnson have shown in [29] that for every circle bundle with even Euler class over a Riemann surface there exists a smooth minimal surface which is a section except over a finite number of points. This result generalises the case of the sphere tangent bundles $T^r S^2$ and their minimal projective planes obtained from the parallel transport vector fields as described in Proposition 5.9.

Singular foliations on spheres, which can be seen as the analogous to the radial vector fields, have been considered by F. G. B. Brito and D. L. Johnson in [17]. These 3-dimensional foliations of the spheres $S^{4m+3}$ consist of all great 3-spheres containing a common great 2-sphere and have the property

**Proposition 5.15** *The 3-dimensional foliations described above minimise the volume among the singular foliations of $S^{4m+3}$ with the same singular locus and limiting behaviour and their volume is a lower bound for the volume of 3-dimensional smooth orientable foliations of $S^{4m+3}$.*

The similar result is also proved for the 7-dimensional foliations of $S^{15}$ consisting of all 7-spheres containing a common great 6-sphere.

# References


1. M. Berger, Du côté de chez Pu. Ann. Sci. Ec. Norm. Sup. **5**, 1–44 (1972)
2. J. Berndt, Real hypersurfaces with constant principal curvatures in complex hyperbolic space. J. Reine Agew. Math. **395**, 132–141 (1989)
3. J. Berndt, L. Vanhecke, L. Verhóczki, Harmonic and minimal unit vector fields on Riemannian symmetric spaces. Illinois J. Math. **47**, 1273–1286 (2003)
4. D.E. Blair, *Riemannian Geometry of Contact and Symplectic Manifolds*. Progress in Mathematics, vol. 203 (Birkhäuser, Boston, 2002)
5. E. Boeckx, L. Vanhecke, Harmonic and minimal vector fields on tangent and unit tangent bundles. Differ. Geom. Appl. **13**, 77–93 (2000)
6. E. Boeckx, L. Vanhecke, Radial vector fields on harmonic manifolds. Bull. Soc. Sci. Math. Roumanie **93**, 181–185 (2000)
7. E. Boeckx, L. Vanhecke, Harmonic and minimal radial vector fields. Acta Math. Hungar. **90**, 317–331 (2001)
8. E. Boeckx, L. Vanhecke, Isoparametric functions and harmonic and minimal unit vector fields, in *Global Differential Geometry: The Mathematical Legacy of Alfred Gray*, ed. by M. Fernández, J.A. Wolf. Contemporary Mathematics Series, vol. 288 (American Mathematical Society, Providence, 2001), pp. 20–31
9. A.A. Borisenko, A.L. Yampol'skii, The sectional curvature of the Sasaki metric of $T_r M^n$. Ukrainskii Geom. Sb. **30**, 10–17 (1987). English translation: Plenum Publishing Corporation, 1990
10. V. Borrelli, Stability of the characteristic vector field of a Sasakian manifold. Soochow J. Math. **30**, 283–292 (2004). Erratum on the article: Stability of the characteristic vector field of a Sasakian manifold. *Soochow J. Math.* **32**, 179–180 (2006)
11. V. Borrelli, O. Gil-Medrano, A critical radius for Hopf vector fields on spheres. Math. Ann. **334**, 731–751 (2006)
12. V. Borrelli, O. Gil-Medrano, Area-minimizing vector fields on round 2-spheres. J. Reine Angew. Math. **640**, 85–99 (2010)
13. V. Borrelli, H. Zoubir, Stability of unit vector fields on quotients of spheres. Differ. Geom. Appl. **28**, 488–499 (2010)
14. V. Borrelli, F.G.B. Brito, O. Gil-Medrano, An energy minimizing family of unit vector fields on odd-dimensional spheres, in *Global Differential Geometry:: The Mathematical Legacy of Alfred Gray*, ed. by M. Fernández, J.A. Wolf. Contemporary Mathematics Series, vol. 288 (American Mathematical Society, Providence, 2001), pp. 273–276
15. V. Borrelli, F.G.B. Brito, O. Gil-Medrano, The infimum of the energy of unit vector fields on odd-dimensional spheres. Ann. Global Anal. Geom. **23**, 129–140 (2003)


 
O. Gil-Medrano, *The Volume of Vector Fields on Riemannian Manifolds*,
Lecture Notes in Mathematics 2336, https://doi.org/10.1007/978-3-031-36857-8

16. F.G.B. Brito, Total Bending of flows with mean curvature correction. Differ. Geom. Appl. **12**, 157–163 (2000)
17. F.G.B. Brito, D.L. Johnson, Volume-minimizing foliations on spheres. Geom. Dedicata **109**, 253–267 (2004)
18. F.G.B. Brito, P. Walczak, Totally geodesic foliations with integrable normal bundles. Bol. Soc. Bras. Mat. **17**, 41–46 (1986)
19. F.G.B. Brito, P. Walczak, On the energy of unit vector fields with isolated singularities. Ann. Math. Polon. **73**, 269–274 (2000)
20. F.G.B. Brito, R. Langevin, H. Rosenberg, Intégrales de courbure sur des variétés feuilletées. J. Differ. Geom. **16**, 19–50 (1981)
21. F.G.B. Brito, P.M. Chacón, A.M. Naveira, On the volume of unit vector fields on spaces of constant sectional curvature. Comment. Math. Helv. **79**, 300–316 (2004)
22. F.G.B. Brito, P.M. Chacón, D.L. Johnson, Unit vector fields on antipodally punctured spheres: big index, big volume. Bull. Soc. Math. France **136**, 147–157 (2008)
23. F.G.B. Brito, A. Gomes, G. Nunes, Energy and volume of vector fields on spherical domains. Pac. J. Math. **257**, 1–7 (2012)
24. F.G.B. Brito, A.O. Gomes, I. Gonçalves, Poincaré index and the volume functional of unit vector fields on punctured spheres. Manuscripta Math. **161**, 487–499 (2020)
25. F.G.B. Brito, I. Gonçalves, A.V. Nicoli, A topological lower bound for the energy of a unit vector field on a closed Euclidean hypersurface. Ann. Polon. Math. **125**, 203–213 (2020)
26. F.G.B. Brito, J. Conrado, I. Gonçalves, A.V. Nicoli, Area minimizing unit vector fields on antipodally punctured unit 2-sphere. C. R. Math. Acad. Sci. Paris **359**, 1225–1232 (2021)
27. F.G.B. Brito, J. Conrado, I. Gonçalves, A.V. Nicoli, G. Nunes, Minimally immersed Klein bottles in the unit tangent bundle of the unit 2-sphere arising from area-minimizing unit vector fields on $S^2 \setminus \{N, S\}$. J. Geom. Anal. **33**, 142, 9pp. (2023)
28. E. Calabi, H. Gluck, What are the best Almost-Complex Structures on the 6-Sphere? Proc. Symp. Pure Math. **54**, 99–106 (1993)
29. P.M. Chacón, D.L. Johnson, Minimal surfaces in circle bundles over Riemann surfaces. Bull. Lond. Math. Soc. **43**, 33–43 (2011)
30. P.M. Chacón, A.M. Naveira, Corrected energy of distributions on Riemannian manifolds. Osaka J. Math. **41**, 97–105 (2004)
31. P.M. Chacón, A.M. Naveira, J.M. Weston, On the energy of distributions, with application to the quaternionic Hopf fibration. Monatshefte für Mathematik **133**, 281–294 (2001)
32. D. DeTurck, H. Gluck, P. Storm, Lipschitz minimality of Hopf fibrations and Hopf vector fields. Algebr. Geom. Topol. **13**, 1369–1412 (2013)
33. P. Dombrowski, On the geometry of the tangent bundle. J. Reine Angew. Math. **210**, 73–88 (1962)
34. S. Dragomir, D. Perrone, *Harmonic Vector Fields: Variational Principles and Differential Geometry*. Progress in Mathematics, vol. 203 (Birkhäuser, Boston, 2002)
35. A.T. Fomenko, Minimal compacta in Riemannian manifolds and Reifensberg's conjecture. Math. USSR **6**, 1037–1066 (1972)
36. I. Fourtzis, M. Markellos, A. Savas-Halilaj, Gauss maps of harmonic and minimal great circle fibrations. Ann. Global Anal. Geom. **63**, 12, 25pp. (2023)
37. S. Gallot, D. Meyer, Opérateur de courbure et laplacien des formes différentielles d'une variété riemannienne. J. Math. Pures Appl. **54**, 259–284 (1975)
38. O. Gil-Medrano, On the volume functional in the manifold of unit vector fields, in *Proceedings of the Workshop on Recent Topics in Differential Geometry, Santiago de Compostela, July 1997*. Publicaciones del Departamento de Geometría y Topología, Universidad de, Santiago, vol. 89 (1998), pp. 155–163
39. O. Gil-Medrano, Relationship between volume and energy of vector fields. Differ. Geom. Appl. **15**, 137–152 (2001)
40. O.Gil-Medrano, Volume and energy of vector fields on spheres. A survey. *Differential Geometry, Valencia, 2001* (World Scientific Publishing, River Edge, 2002), pp. 167–178

41. O. Gil-Medrano, Unit vector fields that are critical points of the volume and of the energy: characterization and examples, in *Complex, Contact and Symmetric Manifolds*. Progress in Mathematics, vol. 234 (Birkhäuser Boston, 2005), pp. 165–186
42. O. Gil-Medrano, Area minimizing projective planes on the projective space of dimension 3 with the Berger metric. C. R. Math. Acad. Sci. Paris **354**, 415–417 (2016)
43. O. Gil-Medrano, Volume minimising unit vector fields on three dimensional space forms of positive curvature. Calc. Var. Partial Differ. Equ. **61**, 66, 8pp. (2022)
44. O. Gil-Medrano, A. Hurtado, Spacelike energy of timelike unit vector fields on a Lorentzian manifold. J. Geom. Phys. **51**, 82–100 (2004)
45. O. Gil-Medrano, A. Hurtado, Volume, energy and generalized energy of unit vector fields on Berger spheres: stability of Hopf vector fields. Proc. Royal Soc. Edinburgh **135A**, 789–813 (2005)
46. O. Gil-Medrano, E. Llinares-Fuster, Second variation of volume and energy of vector fields. Stability of hopf vector fields. Math. Ann. **320**, 531–545 (2001)
47. O. Gil-Medrano, E. Llinares-Fuster, Minimal unit vector fields. Tohoku Math. J. **54**, 71–84 (2002)
48. O. Gil-Medrano, J.C. González-Dávila, L. Vanhecke, Harmonic and minimal invariant unit vector fields on homogeneous Riemannian manifolds. Houston J. Math. **27**, 377–409 (2001)
49. O. Gil-Medrano, J.C. González-Dávila, L. Vanhecke, Harmonicity and minimality of oriented distributions. Isr. J. Math. **143**, 253–279 (2004)
50. H. Gluck, Geodesics in the unit tangent bundle of a round sphere. Ens. Math. **34**, 233–246 (1988)
51. H. Gluck, W. Gu, Volume-preserving great circle flows on the 3-sphere. Geom. Dedicata **88**, 259–282 (2001)
52. H. Gluck, W. Ziller, On the volume of a unit vector field on the three sphere. Comment. Math. Helv. **61**, 177–192 (1986)
53. H. Gluck, F. Morgan, W. Ziller, Calibrated geometries in Grassmann manifolds. Comment. Math. Helv. **64**, 256–268 (1989)
54. J.C. González-Dávila, Harmonicity and minimality of distributions on Riemannian manifolds via the intrinsic torsion. Rev. Mat. Iberoam. **30**, 247–275 (2014)
55. J.C. González-Dávila, L. Vanhecke, Examples of minimal unit vector fields. Ann. Global Anal. Geom. **18**, 385–404 (2000)
56. J.C. González-Dávila, L. Vanhecke, Minimal and harmonic characteristic vector fields on three-dimensional contact metric manifolds. J. Geom. **72**, 65–76 (2001)
57. J.C. González-Dávila, L. Vanhecke, Energy and volume of unit vector fields on three-dimensional Riemannian manifolds. Differ. Geom. Appl. **16**, 225–244 (2002)
58. D.S. Han, J.W. Yim, Unit vector field on spheres which are harmonic maps. Math. Z. **227**, 83–89 (1998)
59. S. Helgason, *Differential Geometry, Lie Groups and Symmetric Spaces*. Pure and Applied Mathematics, vol. 80 (Academic Press, New York, 1978)
60. H. Hopf, Uber die Abbildungen der dreidimensionalen Sphäre auf die Kügelflache. Math. Ann. **104**, 637–665 (1931)
61. A. Hurtado, Stability numbers in K-contact manifolds. Differ. Geom. Appl. **26**, 227–243 (2008)
62. A. Hurtado, Instability of Hopf vector fields on Lorentzian Berger spheres. Isr. J. Math. **177**, 103–124 (2010)
63. T. Ishihara, Harmonic sections of tangent bundles. J. Math. Tokushima Univ. **13**, 23–27 (1979)
64. I. Iwasaki, K. Katase, On the spectrum of the Laplace operator on $\Lambda^*(S^n)$. Proc. Japan Acad. Ser. A Math. Sci. **55**, 141–145 (1979)
65. D.L. Johnson, Volume of flows. Proc. Am. Math. Soc. **104**, 923–932 (1988)
66. D.L. Johnson, P. Smith, Regularity of mass-minimizing one-dimensional foliations, in *Analysis and Geometry in Foliated Manifolds* (Santiago de Compostela, 1994) (World Scientific Publishing, River Edge, 1995), pp. 81–98

67. D.L. Johnson, P. Smith, Partial regularity of mass-minimizing cartesian currents. Ann. Global Anal. Geom. **30**, 239–287 (2006)
68. D.L. Johnson, P. Smith, Regularity of volume-minimizing flows on 3-manifolds. Ann. Global Anal. Geom. **33**, 219–229 (2008)
69. M. Kimura, Real hypersurfaces and complex submanifolds in complex projective spaces. Trans. Am. Math. Soc. **296**, 137–149 (1986)
70. W. Klingenberg, S. Sasaki, On the tangent sphere bundle of a 2-sphere. Tôhoku Math. J. **27**, 49–56 (1975)
71. J.L. Konderak, On sections of fibre bundles which are harmonic maps. Bull. Math. Soc. Sci. Math. Roumanie **42**, 341–352 (1999)
72. O. Kowalski, Curvature of the induced Riemannian metric on the tangent bundle of a Riemannian manifold. J. Reine Angew. Math. **250**, 124–129 (1971)
73. O. Kowalski, M. Sekizawa, On tangent sphere bundles with small or large constant radius. Ann. Global Anal. Geom. **18**, 207–219 (2000)
74. O. Kowalski, F. Tricerri, Riemannian manifolds of dimension $n \leq 4$ admitting a homogeneous structure of class $T_2$, in *Conferenze del Seminario di Matematica University di Bari* , vol. 222 (1987)
75. J. Milnor, Curvature of left invariant metrics on Lie groups. Adv. Math. **21**, 293–329 (1976)
76. O. Nouhaud, Applications harmoniques d'une variété riemannienne dans son fibré tangent. C. R. Acad. Sci. Paris **284**, 815–818 (1977)
77. L. Ornea, L. Vanhecke, Harmonicity and minimality of vector fields and distributions on locally conformal Kähler and hyperkähler manifolds. Bull. Belgian Math. Soc. Simon Stevin **12**, 543–555 (2005)
78. S.L. Pedersen, Volumes of vector fields on spheres. Trans. Am. Math. Soc. **336**, 69–78 (1993)
79. D. Perrone, Harmonic characteristic vector fields on contact metric three-manifolds. Bull. Australian Math. Soc. **67**, 305–315 (2003)
80. D. Perrone, On the volume of unit vector fields on Riemannian trhee-manifolds. C. R. Math. Rep. Acad. Sci. Can. **30**, 11–21 (2008)
81. D. Perrone, Unit vector fields on real space forms which are harmonic maps. Pac. J. Math. **239**, 89–104 (2009)
82. D. Perrone, Unit vector fields of minimum energy on quotients of spheres and stability of the Reeb vector field. Differ. Geom. Appl. **34**, 45–62 (2014)
83. W.A. Poor, *Differential Geometric Structures* (Mc Graw-Hill, New York, 1981)
84. A.G. Reznikov, Lower bounds on volumes of vector fields. Arch. Math. **58**, 509–513 (1992)
85. P. Rukimbira, Criticality of $K$-contact vector fields. J. Geom. Phys. **40**, 209–214 (2002)
86. T. Sakai, *Riemannian Geometry*. Translations of Mathematical Monographs, vol. 149 (American Mathematical Society, Providence, 1996)
87. M. Salvai, On the volume of unit vector fields on a compact semisimple Lie group. J. Lie Theory **13**, 457–464 (2003)
88. S. Sasaki, On the differential geometry of tangent bundles of Riemannian manifolds. Tohoku Math. J. **10**, 338–354 (1958)
89. F. Tricerri, L. Vanhecke, *Homogeneous Structures on Riemannian Manifolds*. London Mathematical Society Lecture Note Series, vol. 83 (Cambridge University Press, Cambridge, 1983)
90. F. Tricerri, L. Vanhecke, Curvature homogeneous Riemannian manifolds. Ann. Sci. École Norm. Sup. **22**, 535–554 (1989)
91. K. Tsukada, L. Vanhecke, Invariant minimal unit vector fields on Lie groups. Period. Math. Hungar. **40**, 123–133 (2000)
92. K. Tsukada, L. Vanhecke, Minimal and harmonic unit vector fields in $G_2(\mathbf{C}^{m+2})$ and its dual space. Monatsh. Math. **130**, 143–154 (2000)
93. K. Tsukada, L. Vanhecke, Minimality and harmonicity for Hopf vector fields. Illinois J. Math. **45**, 441–451 (2001)
94. G. Wiegmink, Total bending of vector fields on Riemannian manifolds. Math. Ann. **303**, 325–344 (1995)

95. J.A. Wolf, *Spaces of Constant Curvature*, 6th edn. (Mc Graw-Hill, New York, 1967). AMS Chelsea Publishing, Providence, 2011
96. J.A. Wolf, A contact structure for odd dimensional spherical space forms. Proc. Am. Math. Soc. **19**, 196 (1968)
97. C.M. Wood, An existence theorem for harmonic sections. Manuscripta Math. **68**, 69–75 (1990)
98. C.M. Wood, A class of harmonic almost-product structures. J. Geom. Phys. **14**, 25–42 (1994)
99. C.M. Wood, Harmonic almost-complex structures. Compos. Math. **99**, 183–212 (1995)
100. C.M. Wood, Harmonic sections of homogeneous fibre bundles. Differ. Geom. Appl. **19**, 193–210 (2003)

*LECTURE NOTES IN MATHEMATICS*

**Editorial Policy**

1. Lecture Notes aim to report new developments in all areas of mathematics and their applications – quickly, informally and at a high level. Mathematical texts analysing new developments in modelling and numerical simulation are welcome.

   Manuscripts should be reasonably self-contained and rounded off. Thus they may, and often will, present not only results of the author but also related work by other people. They may be based on specialised lecture courses. Furthermore, the manuscripts should provide sufficient motivation, examples and applications. This clearly distinguishes Lecture Notes from journal articles or technical reports which normally are very concise. Articles intended for a journal but too long to be accepted by most journals, usually do not have this "lecture notes" character. For similar reasons it is unusual for doctoral theses to be accepted for the Lecture Notes series, though habilitation theses may be appropriate.

2. Besides monographs, multi-author manuscripts resulting from SUMMER SCHOOLS or similar INTENSIVE COURSES are welcome, provided their objective was held to present an active mathematical topic to an audience at the beginning or intermediate graduate level (a list of participants should be provided).

   The resulting manuscript should not be just a collection of course notes, but should require advance planning and coordination among the main lecturers. The subject matter should dictate the structure of the book. This structure should be motivated and explained in a scientific introduction, and the notation, references, index and formulation of results should be, if possible, unified by the editors. Each contribution should have an abstract and an introduction referring to the other contributions. In other words, more preparatory work must go into a multi-authored volume than simply assembling a disparate collection of papers, communicated at the event.

3. Manuscripts should be submitted either online at www.editorialmanager.com/lnm to Springer's mathematics editorial in Heidelberg, or electronically to one of the series editors. Authors should be aware that incomplete or insufficiently close-to-final manuscripts almost always result in longer refereeing times and nevertheless unclear referees' recommendations, making further refereeing of a final draft necessary. The strict minimum amount of material that will be considered should include a detailed outline describing the planned contents of each chapter, a bibliography and several sample chapters. Parallel submission of a manuscript to another publisher while under consideration for LNM is not acceptable and can lead to rejection.

4. In general, **monographs** will be sent out to at least 2 external referees for evaluation.

   A final decision to publish can be made only on the basis of the complete manuscript, however a refereeing process leading to a preliminary decision can be based on a pre-final or incomplete manuscript.

   Volume Editors of **multi-author works** are expected to arrange for the refereeing, to the usual scientific standards, of the individual contributions. If the resulting reports can be forwarded to the LNM Editorial Board, this is very helpful. If no reports are forwarded or if other questions remain unclear in respect of homogeneity etc, the series editors may wish to consult external referees for an overall evaluation of the volume.

5. Manuscripts should in general be submitted in English. Final manuscripts should contain at least 100 pages of mathematical text and should always include

   – a table of contents;
   – an informative introduction, with adequate motivation and perhaps some historical remarks: it should be accessible to a reader not intimately familiar with the topic treated;
   – a subject index: as a rule this is genuinely helpful for the reader.
   – For evaluation purposes, manuscripts should be submitted as pdf files.

6. Careful preparation of the manuscripts will help keep production time short besides ensuring satisfactory appearance of the finished book in print and online. After acceptance of the manuscript authors will be asked to prepare the final LaTeX source files (see LaTeX templates online: https://www.springer.com/gb/authors-editors/book-authors-editors/manuscriptpreparation/5636) plus the corresponding pdf- or zipped ps-file. The LaTeX source files are essential for producing the full-text online version of the book, see http://link.springer.com/bookseries/304 for the existing online volumes of LNM). The technical production of a Lecture Notes volume takes approximately 12 weeks. Additional instructions, if necessary, are available on request from lnm@springer.com.

7. Authors receive a total of 30 free copies of their volume and free access to their book on SpringerLink, but no royalties. They are entitled to a discount of 33.3 % on the price of Springer books purchased for their personal use, if ordering directly from Springer.

8. Commitment to publish is made by a *Publishing Agreement*; contributing authors of multiauthor books are requested to sign a *Consent to Publish form*. Springer-Verlag registers the copyright for each volume. Authors are free to reuse material contained in their LNM volumes in later publications: a brief written (or e-mail) request for formal permission is sufficient.

**Addresses:**

Professor Jean-Michel Morel, CMLA, École Normale Supérieure de Cachan, France
E-mail: moreljeanmichel@gmail.com

Professor Bernard Teissier, Equipe Géométrie et Dynamique,
Institut de Mathématiques de Jussieu – Paris Rive Gauche, Paris, France
E-mail: bernard.teissier@imj-prg.fr

Springer: Ute McCrory, Mathematics, Heidelberg, Germany,
E-mail: lnm@springer.com

Printed in the United States
by Baker & Taylor Publisher Services